U.S. Space Force

TD Barnes

Published by TD Barnes, 2024.

U.S. SPACE FORCE

First edition. October 16, 2024.

ISBN: 979-8227994387

Written by TD Barnes.

Table of Contents

US Space Force
Guardians of the High Frontier

By: TD Barnes

Copyright 2024 TD Barnes

Foreword

In the vast expanse beyond our atmosphere, space has evolved from a mysterious frontier to the next theater of national security. Once the realm of exploration and scientific discovery, it is now a domain where the future of defense is being shaped. The establishment of the United States Space Force marks a pivotal moment in military history, a recognition of the critical importance of space in maintaining national security and global stability.

The US Space Force - Guardians of the High Frontier takes you on a journey from the early days of space exploration, born out of Cold War rivalries, to the creation of the Space Force in 2019, and its vital role in safeguarding the American way of life. This book is not just a chronicle of the Space Force's inception; it is an exploration of the geopolitical forces and technological advancements that demanded its formation.

In these pages, you will delve into the Cold War era, where space became the ultimate battleground for superpowers, and where the 1947 National Security Act laid the groundwork for today's military structure. You will examine the challenges posed by the Russian Space Forces and China's People's Liberation Army, both of which continue to push the boundaries of space militarization. From NASA's pioneering achievements to the clandestine operations of the National Reconnaissance Office, every chapter highlights the intricate web of organizations and strategies that have shaped the modern space race.

Through chapters detailing the role of the Department of Defense, the contributions of the Army and Navy's space commands, and the rise of anti-satellite weapons, this book paints a comprehensive picture of how space has transitioned from a distant dream to a strategic imperative. The culmination of this narrative is the creation of the U.S. Space Force, a bold new branch tasked with defending America's interests in the high frontier.

As you explore the history and future of space defense, you will gain a deeper understanding of how the Space Force is not just a new branch of the military—it is a guardian of our future. The Space Force stands as a testament to the foresight of leaders who understood that, in the boundless expanse of space, the stakes are higher than ever before.

In The US Space Force - Guardians of the High Frontier, you will find a comprehensive and insightful guide to the forces, both historical and modern, that have driven the evolution of military space operations. This book is for anyone who seeks to understand the strategic importance of space in the 21st century and the critical role the Space Force plays in securing the United States' position as the leader in this final frontier.

The Guardians of the High Frontier are not just defending satellites—they are defending our future.

Chapter 1 - Space Exploration: From the Cold War to the Modern Era

Space exploration has opened a new frontier by using astronomy and space technology to investigate outer space. While astronomers have long explored the universe through telescopes, the physical exploration of space began with space technology in the mid-20th century, enabling uncrewed robotic probes and human spaceflight to venture beyond Earth's atmosphere. Space exploration continues to drive scientific discovery, international cooperation, and technological advancement.

Space exploration has been justified through various economic, scientific, and existential arguments. One of the primary reasons for government investment in space programs, such as NASA and Roscosmos, is the economic benefit they generate. Economic analyses consistently show that these programs provide returns far exceeding their costs. NASA, for instance, has produced numerous technological innovations, known as "spin-offs," that have commercial and societal applications far beyond the space industry. These range from advancements in healthcare technologies to innovations in communication and transportation systems.

Proponents of space exploration also argue that it holds the potential for future economic benefits through resource extraction. Asteroids, in particular, are rich in valuable minerals and metals, with estimates suggesting that they contain trillions of dollars worth of resources. The ability to mine these celestial bodies could lead to entirely new industries in space. Furthermore, space exploration inspires future generations, particularly in science, technology, engineering, and mathematics (STEM). This inspirational effect is crucial in driving innovation and maintaining a competitive edge in the global economy.

In addition to these tangible benefits, space exploration is often seen as a necessity for the survival of humanity. Stephen Hawking famously warned that remaining confined to Earth presents significant risks, from natural resource depletion to global catastrophes like nuclear war or pandemics. He and other advocates argue that colonizing other planets is the only way to ensure humanity's long-term survival. Arthur C. Clarke, in his 1950 work Interplanetary Flight, echoed these sentiments, presenting space expansion as humanity's only alternative to stagnation and eventual extinction. This philosophy, known as the Von Braun Paradigm, envisioned a future where humanity would build reusable spacecraft, establish permanent space stations, and eventually colonize the Moon and Mars.

Wernher von Braun's vision for space exploration laid the groundwork for much of NASA's human spaceflight efforts. His plan involved five key steps: developing multi-stage rockets to place satellites, animals, and humans into space; creating reusable spacecraft to lower the cost of space access; constructing a permanently occupied space station; sending humans to the Moon with the intent of building lunar bases; and, finally, assembling and fueling spacecraft in Earth orbit for manned missions to Mars. Though these steps were not completed in order—Apollo missions landed humans on the Moon before the space shuttle program began—the Von Braun Paradigm has remained a central guiding philosophy in space exploration efforts.

Public sentiment toward space exploration has largely remained positive. Polls, such as one conducted by the Associated Press in 2003, found that 71% of Americans viewed the space program as "a good investment." This support reflects the belief that space exploration is not only a source of national pride but also an investment in scientific advancement and future technological breakthroughs. NASA and other space agencies have produced public service announcements to reinforce the value of space exploration, highlighting its potential benefits to humanity.

Spaceflight: The Means of Exploration

Spaceflight is the foundation of space exploration, enabling spacecraft to launch into orbit and beyond. This highly technical endeavor involves overcoming Earth's gravitational pull with rockets, which provide the necessary thrust to propel spacecraft into space. Once a spacecraft achieves orbit, its motion is governed by the principles of astrodynamics, which cover both powered and unpowered flight through space.

Spaceflight is employed for exploration and commercial activities such as satellite telecommunications and emerging industries like space tourism. Satellites launched into space serve a variety of non-commercial purposes as well, including Earth observation, weather monitoring, and reconnaissance. Some spacecraft, like space observatories, remain in space indefinitely, while others are designed to return to Earth or land on planetary or lunar surfaces. Both crewed and uncrewed spaceflight is essential for conducting experiments in space, exploring new frontiers, and expanding humanity's presence beyond Earth.

Space advocacy often appeals to humans' inherent drive to explore the unknown. This idea of exploration as part of human nature underscores many arguments regarding space exploration. Humans have ventured into unknown territories for knowledge, resources, and new opportunities throughout history. Today, space represents the ultimate frontier, offering boundless potential for discovery and expansion. As technological capabilities advance, spaceflight and exploration will continue to push the boundaries of human achievement, fulfilling a deep-seated desire to explore the cosmos.

In essence, space exploration is viewed not only as a scientific or economic endeavor but also as an existential necessity, ensuring the survival and advancement of humanity. Whether for the practical benefits of new technologies and resources or the philosophical pursuit of survival and expansion, space exploration continues to captivate and inspire, driving efforts to explore the universe's farthest reaches.

The observation of celestial objects has roots in antiquity. Still, it was only with the development of powerful rockets in the mid-20th century that humans gained the ability to explore space physically. This transformative moment in history occurred during the height of the Cold War when geopolitical competition between the Soviet Union and the United States fueled a rapid acceleration in space technology.

The launch of the first human-made object into Earth's orbit—Sputnik 1 by the Soviet Union on October 4, 1957—marked the official beginning of the "Space Race." This intense rivalry aimed to demonstrate technological superiority, with both superpowers vying for global dominance. The United States responded by founding NASA (the National Aeronautics and Space Administration) in 1958, focusing its efforts on space exploration to assert national prestige and defense capabilities.

Throughout the late 1950s and 1960s, the Soviet Union achieved many significant firsts in space exploration, setting the pace in this international race. Key milestones included:

Sputnik 1: The first artificial satellite, launched by the Soviet Union in 1957, demonstrated the ability to send objects into orbit.

Laika: The first living creature in space, a dog sent aboard Sputnik 2 in 1957.

Yuri Gagarin: The first human to travel into space and orbit Earth aboard Vostok 1 on April 12, 1961.

Alexei Leonov: The first person to perform a spacewalk, exiting his spacecraft during the Voskhod 2 mission on March 18, 1965.

Luna 9: The first successful landing of a spacecraft on another celestial body, the Moon, achieved by the Soviet Union in 1966.

Salyut 1: The first space station, launched by the Soviet Union in 1971, laid the groundwork for long-duration human habitation in space.

Despite the Soviet Union's initial dominance, the United States achieved arguably the most iconic moment in space exploration history. On July 20, 1969, NASA's Apollo 11 mission successfully landed the first humans on the Moon. Astronauts Neil Armstrong and Buzz Aldrin took humankind's first steps on another world, fulfilling

President John F. Kennedy's ambitious goal of landing a man on the Moon and returning him safely to Earth before the decade's end. This achievement not only underscored American ingenuity but also shifted the direction of space exploration from competition to cooperation.

By the 1970s, the intense rivalry between the Soviet Union and the United States began to soften, and both nations realized the value of collaboration in space. The Apollo-Soyuz Test Project marked this shift in priorities in 1975, where American and Soviet spacecraft docked in orbit, symbolizing a new era of cooperation.

NASA's development of reusable spacecraft became a central focus in the following decades. In 1981, the United States launched the Space Shuttle program, revolutionizing space travel by enabling multiple missions with the same spacecraft. The Space Shuttle fleet would go on to support various missions, including the deployment of satellites, servicing of the Hubble Space Telescope, and construction of the International Space Station (ISS).

The ISS, an unparalleled example of international collaboration, began construction in 1998 and became fully operational by the early 2000s. This orbiting laboratory brought together contributions from space agencies across the globe, including NASA, Roscosmos (Russia), JAXA (Japan), ESA (Europe), and CSA (Canada). The ISS provided a platform for long-term human habitation and scientific research, fostering partnerships that continue to this day.

In the 21st century, space exploration entered a new phase with the rise of private companies and the expansion of space programs in countries outside the Cold War duopoly. The 2010s witnessed the advent of the private space industry, led by companies such as SpaceX, Blue Origin, and Rocket Lab, which developed commercial launch vehicles, reusable rockets, and satellite systems. These developments have drastically reduced the cost of access to space, enabling new opportunities for exploration, research, and business ventures.

Meanwhile, emerging space powers like China and India have made significant strides in space exploration. China's Shenzhou crewed spaceflight program has seen several successful missions, including establishing China's space station, Tiangong, and launching the Chang'e lunar program, which includes ambitious plans for Moon exploration. India's Chandrayaan program has also achieved notable successes, including discovering water on the Moon's surface, while Japan and Europe continue to plan for future crewed missions.

The 2020s have ushered in two major international initiatives: the US-led Artemis Program and the China-led International Lunar Research Station project. The Artemis Program, named after the sister of Apollo in Greek mythology, aims to return humans to the Moon by 2025 and establish a sustainable lunar presence by constructing the Lunar Gateway and Artemis Base Camp. China's lunar ambitions focus on building a permanent outpost on the Moon in partnership with Russia and other international collaborators.

As space exploration advances, humanity looks beyond the Moon to more distant destinations, with Mars emerging as the next great frontier. Robotic missions, such as NASA's Perseverance rover and China's Tianwen-1, have already begun exploring the Martian surface, gathering data and paving the way for future crewed missions to the Red Planet.

The future of space exploration holds immense promise, from uncovering the mysteries of our solar system to expanding humanity's presence in the cosmos. As nations and private entities continue to push the boundaries of space travel, cooperation and innovation will shape the next era of discovery, ensuring that space remains a domain of peace and opportunity for all.

The history of space exploration is a remarkable journey, spanning centuries of astronomical discoveries and technological advancements that have transformed our understanding of the universe. From the invention of the first telescopes to modern space stations, humans have continuously pushed the boundaries of what is possible in exploring space.

The exploration of space began with the invention of the telescope. In 1608, Dutch eyeglass maker Hans Lippershey is credited with creating the first telescope. However, it was Galileo Galilei who, in 1609, first used a telescope for astronomical purposes. Galileo's observations led to groundbreaking discoveries, including the mountains of the Moon, the phases of Venus, and the four largest moons of Jupiter (now known as the Galilean moons). In 1668, Isaac Newton built the first fully functional reflecting telescope, significantly improving Galileo's design due to its superior optical capabilities.

These early telescopes paved the way for centuries of astronomical discoveries, revealing the complexities of the Solar System and beyond. Astronomers soon identified Saturn's rings, new comets, asteroids, and distant planets such as Uranus (discovered by William Herschel in 1781) and Neptune (discovered by Johann Galle in 1846). This period of discovery laid the foundation for modern space science and exploration.

While ground-based telescopes revolutionized astronomy, the next leap forward came with space-based observatories. In 1968, the Orbiting Astronomical Observatory 2 became the first space telescope, allowing scientists to observe the universe without interference from Earth's atmosphere. However, the most significant milestone in space-based astronomy was the launch of the Hubble Space Telescope in 1990. Hubble's detailed observations have contributed to discoveries such as the expansion rate of the universe and the existence of exoplanets—planets orbiting stars outside our Solar System.

As of December 2022, astronomers have confirmed the existence of over 5,000 exoplanets, providing insights into the potential for life beyond Earth. The Milky Way alone is estimated to contain between 100 and 400 billion stars and over 100 billion planets. Observations of distant galaxies have expanded our view of the cosmos, with current estimates suggesting there are at least two trillion galaxies in the observable universe.

The journey into outer space began during World War II with German engineers' development of the V-2 rocket. On June 20, 1944, a V-2 rocket, launched from the Peenemünde Army Research Center, became the first human-made object to reach space, achieving an altitude of 176 kilometers—well beyond the Kármán line, the boundary of space. Although the rocket did not reach orbital velocity, it began human efforts to explore outer space.

The first successful orbital launch came on October 4, 1957, when the Soviet Union launched Sputnik 1, the world's first artificial satellite. Weighing 83 kilograms, Sputnik 1 orbited Earth at approximately 250 kilometers and transmitted radio signals back to Earth, providing valuable data about the ionosphere.

Less than four years later, on April 12, 1961, Soviet cosmonaut Yuri Gagarin made history as the first human to orbit Earth aboard Vostok 1. Gagarin's 108-minute flight marked the dawn of human spaceflight and demonstrated the Soviet Union's leadership in space exploration during the early years of the Cold War.

Following the success of early space missions, both the United States and the Soviet Union aimed to explore other celestial bodies, particularly the Moon. The Soviet Luna 2 mission became the first spacecraft to reach the Moon in 1959, followed by Luna 9, which achieved the first soft landing on the lunar surface in 1966. The Apollo 11 mission, however, remains the most iconic achievement of space exploration. On July 20, 1969, Neil Armstrong and Buzz Aldrin became the first humans to walk on the Moon, fulfilling President John F. Kennedy's goal and solidifying the United States' position in space exploration.

The exploration of other planets soon followed. Venera 1, launched by the Soviet Union in 1961, was the first spacecraft to fly by Venus, although it lost contact before returning data. NASA's Mariner 2 conducted the first successful flyby of Venus in 1962. Over the following decades, robotic probes flew by every planet in the Solar System. Notable achievements include Mariner 4's 1965 flyby of Mars, Pioneer 10's 1973 flyby of Jupiter, and Voyager 2's encounters with Uranus (1986) and Neptune (1989). In 2015, NASA's New Horizons probe made the first close flyby of Pluto, expanding our understanding of the Solar System's outer reaches.

Mars exploration has been a central focus of robotic space missions. In 1971, the Soviet Mars 3 became the first spacecraft to achieve a soft landing on Mars, although it transmitted data for only 20 seconds. NASA's Viking 1, which landed on Mars in 1976, conducted the first long-term mission, transmitting data for over six years. These missions laid the groundwork for future exploration, including NASA's Perseverance rover, which continues to explore the Martian surface today.

Space stations represent a significant advancement in human space exploration, providing a platform for long-term habitation and scientific space research. The first space station, Salyut 1, was launched by the Soviet Union on April 19, 1971. Over the decades, space stations have evolved from short-term experimental platforms to sophisticated laboratories. The International Space Station (ISS), launched in 1998, is the largest and longest-operating space station. It has been continuously inhabited since 2000, representing a monumental achievement in international cooperation between NASA, Roscosmos, and other space agencies.

In recent years, China has developed its own space station, Tiangong, which became fully operational and crewed in the early 2020s. Tiangong is part of China's ambitious space program, which includes plans for lunar exploration and potential human missions to Mars.

On August 25, 2012, Voyager 1 became the first human-made object to enter interstellar space, a milestone in space exploration. After launching in 1977, Voyager 1 journeyed through the outer Solar System, eventually passing the heliopause—the boundary where the Sun's influence wanes—at 121 astronomical units (AU) from Earth. This marked humanity's first venture into the vast, uncharted territory beyond the solar wind, entering the interstellar medium, a region filled with particles and radiation from other stars. As of November 26, 2022, Voyager 1 was approximately 159 AU (23.8 billion kilometers or 14.8 billion miles) from Earth, making it the most distant human-made object.

In 1970, the crew of Apollo 13 inadvertently set a record during their mission when they traveled farther from Earth than any humans had before—or since. After a catastrophic explosion aboard their spacecraft, the crew had to abort their lunar landing and loop around the Moon for a safe return. During this emergency course correction, Apollo 13 passed the far side of the Moon at an altitude of 254 kilometers (158 miles) above the lunar surface. At its farthest point, they were 400,171 kilometers (248,655 miles) from Earth. This remains the farthest distance humans have ever traveled from their home planet.

Beginning in the mid-20th century, humanity expanded its reach beyond Earth's orbit, first by sending robotic probes and later humans to the Moon. Uncrewed spacecraft were dispatched across the Solar System to explore planets, moons, and the Sun. By the 21st century, spacecraft had orbited or flown by all of the Solar System's planets, including Mercury, Venus, Mars, Jupiter, and Saturn. Among the most distant travelers, Voyager 1 and Voyager 2 both crossed 100 AU, continuing their mission beyond the Sun's heliosphere and into interstellar space. These missions allowed scientists to examine the Solar System's boundary region, known as the heliosheath, providing valuable insights into the Sun's influence and the nature of interstellar space.

The Sun has been a central focus of space exploration, particularly in understanding its solar wind, magnetic field, and radiation—phenomena that impact Earth's space weather, power grids, and satellites. Space-based observations of the Sun, beginning with the Apollo Telescope Mount, have been crucial in revealing the nature of solar activity. One of the most ambitious missions is the Parker Solar Probe, launched in 2018, which will approach the Sun within 1/9th the distance of Mercury's orbit. This mission aims to study the solar corona and better understand the solar wind and its effects on the Solar System.

Mercury, the smallest and least explored terrestrial planet, has only been visited by two spacecraft—Mariner 10 in 1974-75 and MESSENGER, which orbited the planet from 2011 to 2015. Due to its proximity to the Sun, reaching Mercury requires significant energy, and maintaining a stable orbit is challenging. The upcoming BepiColombo mission, a joint venture between the European Space Agency and the Japan Aerospace Exploration Agency, is scheduled to arrive in 2025. It will send two probes to complement MESSENGER's findings and further investigate Mercury's composition, magnetic field, and geological history.

Venus, Earth's "sister planet," was the first target for interplanetary exploration. In 1961, the Soviet Union launched Venera 1, and the U.S. followed with Mariner 2 in 1962, the first successful flyby mission. Despite Venus's hostile environment, with surface temperatures exceeding 450°C (842°F) and crushing atmospheric pressure, the Soviets landed several probes. Venera 7, in 1970, was the first spacecraft to land on Venus and transmit data from its surface. The Venera program continued into the 1980s, with landers providing the first images of Venus's surface and radar-equipped orbiters mapping its terrain beneath the thick cloud cover. Today, Venus remains a key subject for future exploration, with plans to study its atmospheric evolution and volcanic activity.

Space exploration has also revolutionized our understanding of Earth. The first space-based observations of Earth began with the launch of TIROS-1 in 1960, providing humanity's first images of the planet from space. These early missions led to significant discoveries, such as the Van Allen radiation belts, zones of charged particles trapped by Earth's magnetic field that pose a significant challenge for human space exploration. Satellites have since been used to monitor environmental changes, including discovering ozone holes, and to uncover geological formations and archaeological sites that are invisible from the ground. Earth observation satellites are crucial in climate monitoring, disaster response, and natural resource management.

The Moon was humanity's first target for space exploration beyond Earth. In 1959, the Soviet Luna 3 probe captured the first images of the far side of the Moon, a region never before seen by humans. The U.S. followed with the Ranger and Surveyor programs, culminating in the crewed Apollo missions. In 1969, Apollo 11 marked a historic achievement when Neil Armstrong and Buzz Aldrin became the first humans to set foot on the lunar surface. Six Apollo missions successfully landed on the Moon, with the last, Apollo 17, in 1972. These missions provided invaluable scientific data, including lunar rock samples and measurements of the Moon's gravitational field.

In recent years, lunar exploration has been resurgent, with China's Chang'e program making significant strides. In 2019, Chang'e 4 became the first mission to land on the far side of the Moon, and Chang'e 6 aims to return samples from this region by 2024. India also made history with its Chandrayaan-3 mission, which successfully landed near the Moon's south pole in 2023, a region of great interest for future exploration due to the presence of water ice. The upcoming Artemis program, led by NASA, plans to return humans to the Moon for the first time since Apollo. Artemis 2 is set to perform a crewed flyby in 2025, with Artemis 3 targeting a lunar landing by 2026.

Mars has long been a prime target for space exploration, captivating scientists and engineers alike. Since the 1960s, numerous robotic missions have been launched by countries including the United States, the Soviet Union (later Russia), Europe, Japan, and India. These missions have included orbiters, landers, and rovers, each aimed at uncovering the mysteries of the Red Planet. The primary goals of Mars exploration have been to understand its current environment, investigate its geological history, and determine whether life ever existed there. The insights gained from studying Mars also offer valuable comparisons to Earth's past, present, and potential future.

Despite the scientific importance, Mars exploration has been fraught with challenges. Nearly two-thirds of the missions sent to Mars have failed, earning it a reputation among space engineers as being cursed by what is humorously called "The Great Galactic Ghoul." This high failure rate is largely due to the complexity of interplanetary travel and the difficulties posed by Mars' thin atmosphere and harsh surface conditions. However, there have also been notable successes. In 2014, India became the first nation to succeed in reaching Mars on its maiden attempt with the Mars Orbiter Mission (MOM), one of the least expensive interplanetary missions in history at a cost of only $73 million. More recently, the United Arab Emirates launched the Emirates Mars Mission, also known as the "Hope Probe," which successfully entered Mars' orbit in February 2021 to study the planet's atmosphere in unprecedented detail.

Phobos, one of Mars' two moons, has also been a subject of exploration interest. The Russian mission Fobos-Grunt, launched in 2011, aimed to land on Phobos and return samples to Earth. Unfortunately, the spacecraft encountered a failure shortly after launch, leaving it stranded in low Earth orbit. Had the mission succeeded, it would have studied Phobos as a potential stopover for future Mars-bound missions, highlighting the moon's possible role as a trans-shipment point for spacecraft traveling to and from Mars.

Before the advent of space travel, asteroids were little more than specks of light in even the largest telescopes. However, in recent decades, several space missions have unlocked the secrets of these small, rocky bodies. NASA's Galileo spacecraft was the first to visit asteroids, flying past 951 Gaspra in 1991 and 243 Ida in 1993 while en route to Jupiter. The first landing on an asteroid came in 2000, when the NEAR Shoemaker probe touched down on 433 Eros after a thorough orbital survey.

Japan's Hayabusa mission made history by returning samples from the near-Earth asteroid Itokawa in 2010. This pioneering mission demonstrated the ability to bring asteroid material back to Earth for study, providing valuable insights into the formation of the Solar System. NASA's Dawn mission further expanded our understanding by visiting two of the largest bodies in the asteroid belt, 4 Vesta and the dwarf planet Ceres, between 2011 and 2015.

Jupiter, the largest planet in the Solar System, has been explored solely through robotic missions. Since 1973, NASA has sent several spacecraft to conduct flybys of the gas giant, gathering valuable data about its atmosphere, moons, and magnetosphere. The Pioneer and Voyager missions are notable among these, as they provided the first detailed observations of the planet. The Galileo spacecraft became the first to orbit Jupiter, conducting an in-depth study from 1995 to 2003. More recently, the Juno spacecraft entered Jupiter's orbit in 2016 to study its magnetic fields and auroras, aiming to understand the planet's formation and structure.

Despite these successful missions, landing on Jupiter remains impossible due to its lack of a solid surface and immense gravitational forces. With 95 known moons, Jupiter also presents a wealth of exploration opportunities, particularly on its largest moon, Ganymede, and its ocean-bearing moon, Europa, both prime targets for future missions.

Saturn, known for its stunning ring system, has been explored through a series of unmanned spacecraft. Pioneer 11 first flew past Saturn in 1979, followed by Voyager 1 in 1980 and Voyager 2 in 1981. These missions provided the first close-up images of Saturn's rings and moons. However, it was the Cassini-Huygens mission, launched in 1997, that provided the most detailed exploration of Saturn. Cassini orbited Saturn from 2004 to 2017, studying its atmosphere, rings, and moons, while the Huygens probe landed on Titan, Saturn's largest moon, in 2005. This landing marked the first and only time a spacecraft has touched down on an object in the outer Solar System. Titan's thick atmosphere and the discovery of liquid methane lakes make it a particularly intriguing world, with potential for future exploration.

Uranus remains one of the least explored planets in the Solar System, with only one spacecraft, Voyager 2, having visited it during a flyby in 1986. Voyager 2 revealed a planet with a remarkably featureless appearance, lacking the prominent atmospheric activity seen on Jupiter and Saturn. However, the spacecraft discovered a unique magnetosphere and provided detailed images of Uranus's ring system and moons. Uranus's extreme axial tilt, which causes one of its poles to face the Sun for extended periods, adds to the planet's scientific intrigue. However, no follow-up missions are currently planned.

Neptune was also visited by Voyager 2, in 1989, marking the first and only close encounter with the distant ice giant. Contrary to expectations, Neptune displayed dynamic atmospheric activity, including banding and the Great Dark Spot—a massive storm system similar to Jupiter's Great Red Spot. Voyager 2 also discovered Neptune's faint rings and several moons, including Proteus, the last large moon to be discovered in the Solar System. Neptune's largest moon, Triton, is believed to be a captured Kuiper Belt object, making it a fascinating target for future exploration. While a Neptune orbiter has been discussed, no concrete plans for such a mission have yet materialized.

Pluto, once classified as the ninth planet, was first visited by NASA's New Horizons mission in 2015, providing the first detailed images of the dwarf planet and its largest moon, Charon. Launched in 2006, New Horizons traveled over 4.8 billion kilometers (3 billion miles) before its close encounter with Pluto. The spacecraft's observations revealed a surprisingly active world with towering ice mountains and vast nitrogen plains. After its Pluto flyby, New Horizons continued on to study a second Kuiper Belt object, Arrokoth, in 2019, marking the first time a spacecraft visited such a distant object in the outer reaches of the Solar System.

Comet exploration has yielded some of the most dramatic space science achievements. NASA's Deep Impact mission, which deliberately crashed a probe into Comet 9P/Tempel in 2005, provided unprecedented insights into the composition of a comet's nucleus. The European Space Agency's Rosetta mission achieved another milestone when its Philae lander touched on Comet 67P/Churyumov–Gerasimenko in 2014. These missions and others like NASA's Stardust, which returned samples from the tail of Comet Wild 2, have helped scientists understand these ancient remnants of the early Solar System.

Deep space exploration involves studying and physically exploring distant regions of outer space, beyond Earth and its immediate vicinity. This field, which merges astronomy, astronautics, and space technology, is conducted through human and robotic spacecraft missions. The goal is to extend human knowledge about the universe's farthest reaches, exploring planets, moons, asteroids, and other celestial objects. As of the 21st century, robotic spacecraft have visited every planet in the Solar System and have ventured beyond to study phenomena such as interstellar space and distant galaxies.

Some of the most promising technologies being considered for future deep space exploration include antimatter propulsion, nuclear-powered spacecraft, and beamed propulsion systems. Beamed propulsion, which uses focused energy beams to propel spacecraft, is currently seen as one of the most feasible options, as it relies on well-understood physics and existing technology being developed for other purposes. This propulsion method could significantly reduce travel times to distant destinations, making deep space missions more practical.

The future of space exploration is driven by bold initiatives aimed at extending humanity's presence beyond Earth. Research into advanced propulsion technologies and life-support systems continues progressing, focusing on enabling long-duration missions to distant planets, moons, and even stars. One such initiative is the Breakthrough Starshot project, which aims to develop a fleet of tiny light sail spacecraft capable of traveling to the Alpha Centauri star system, located 4.37 light-years away. Announced in 2016 by a team that included Yuri Milner, Stephen Hawking, and Mark Zuckerberg, the project envisions sending these ultra-light spacecraft at a fraction of the speed of light, using laser propulsion, to conduct the first interstellar reconnaissance of a neighboring star system.

Asteroids have become a key focus of space exploration, with some experts proposing them as waypoints for missions to Mars. This concept involves using asteroids as stepping stones to extend flight durations and explore deeper into space, while also providing shelter from harmful cosmic radiation. For such an approach to be viable, several steps must be taken: conducting a comprehensive survey to identify suitable asteroids near Earth, improving the capability for extended missions, and developing advanced robotic tools for astronauts to explore asteroids, regardless of their size or spin.

Asteroids also offer scientific value, containing materials that could reveal much about the early Solar System. Missions like NASA's OSIRIS-REx and Japan's Hayabusa2 have successfully visited and collected samples from asteroids, demonstrating the potential for future asteroid-based exploration.

The James Webb Space Telescope (JWST), launched in December 2021, represents the next generation of space telescopes, designed to surpass the capabilities of the Hubble Space Telescope. JWST is equipped with a larger mirror—6.5 meters in diameter compared to Hubble's 2.4 meters—and operates primarily in the infrared spectrum, allowing it to peer deeper into the universe and observe objects that are too distant or faint for Hubble to detect. JWST will shed light on the formation of the earliest stars and galaxies by observing high-redshift galaxies, offering unparalleled insights into the universe's origins.

JWST's mission also includes studying the formation of stars and planets, direct imaging of exoplanets, and investigating distant supernovae. Positioned at the Earth-Sun L2 Lagrangian point, the telescope is protected by a massive sunshield, ensuring it remains cold enough to capture faint infrared signals. This groundbreaking mission is poised to revolutionize our understanding of the cosmos, opening up new avenues for exploration and discovery.

NASA's Artemis program represents humanity's return to the Moon, with plans to land the first woman and the next man on the lunar surface. Set to target the Moon's south pole by 2025, Artemis is the next step toward establishing a sustainable presence on the Moon, which will serve as a springboard for future missions to Mars. The Artemis missions build on the technologies developed during the Apollo program but aim to create a long-term lunar economy, driven by international collaboration and commercial partnerships.

Artemis relies on several key components, including the Space Launch System (SLS), the Orion spacecraft, and the Lunar Gateway—a space station that will orbit the Moon and provide a staging point for missions to the lunar surface and beyond. NASA's partnerships with international space agencies, such as the European Space Agency (ESA), and commercial entities are crucial to the program's success. In 2022, Artemis I, an uncrewed mission, successfully tested the SLS and Orion spacecraft in preparation for future crewed missions. Artemis II, a crewed lunar flyby, is scheduled for 2025, with Artemis III planned to be the first mission to return humans to the Moon's surface since Apollo 17.

Additionally, NASA is working on the Commercial Lunar Payload Services (CLPS) initiative, which aims to land robotic payloads on the Moon to support science and technology development. As of 2024, the first CLPS lander had successfully reached the lunar surface, marking the United States' first Moon landing since the Apollo era. The Artemis program is not only a stepping stone to Mars but also an essential part of creating a robust space exploration infrastructure that will enable humanity to explore deep space for decades to come.

The commercialization of space began with the launch of private satellites by government space agencies like NASA. This laid the foundation for industries such as satellite navigation, satellite television, and satellite radio, which have since become integral to global communication and navigation systems. However, the next phase of commercialization focused on human spaceflight. While space agencies like NASA initially led human missions to space, reusable spacecraft were the next major innovation in making space travel more accessible and cost-effective. The idea of reusable vehicles was championed by prominent figures such as Buzz Aldrin, who saw them as the key to affordable space travel. Aldrin argued that passenger space travel had the potential to become a large market, justifying the development of reusable launch vehicles.

The introduction of reusable spacecraft, exemplified by NASA's Space Shuttle program, marked a significant milestone. It opened the door to space tourism, where individuals could travel to space for personal enjoyment. Private companies, such as SpaceX and Blue Origin, have taken this vision forward, developing reusable rockets and spacecraft to support space tourism and commercial ventures like satellite launches and space station resupply missions. Companies like Axiom Space and Bigelow Aerospace are working to create commercial space stations, further transforming the space industry into a marketplace of private enterprise.

Astrobiology is the interdisciplinary study of life in the universe, focusing on the origins, distribution, and evolution of life beyond Earth. This field combines elements of astronomy, biology, and geology to explore the potential for life on other planets and moons. Astrobiologists study both the chemical and biological factors that could allow life to exist in extreme environments. In the search for extraterrestrial life, certain locations in the Solar System are considered prime candidates, including the icy moons Enceladus, Europa, Titan, and the planet Mars.

Astrobiology also considers the possibility of chemically distinct life from life on Earth. This includes forms of life that might be based on different chemical elements or that thrive in environments considered inhospitable by Earth standards. With each new mission to these celestial bodies, astrobiologists hope to gather evidence to help answer one of humanity's oldest questions: are we alone in the universe?

Human spaceflight, a critical component of space exploration, has been ongoing for over six decades. The longest continuous human space occupation is aboard the International Space Station (ISS), which has been operational for over 23 years. The ISS serves as both a laboratory for scientific experiments and a platform for studying the effects of long-term space habitation. The record for the longest single spaceflight is held by Russian cosmonaut Valeri Polyakov, who spent nearly 438 consecutive days aboard the Mir space station. These missions provide valuable data on how prolonged exposure to space affects the human body, including bone density loss, muscle atrophy, and radiation exposure.

The health risks associated with spaceflight are numerous. Space motion sickness, caused by disrupting the neurovestibular system in microgravity, affects nearly all astronauts in the first few days of a mission. Long-term effects include vision problems caused by increased pressure on the eyes, cardiovascular deconditioning, and weakened immune systems. Radiation poses a serious threat, especially during deep space missions, where astronauts are exposed to higher levels of solar and cosmic radiation outside Earth's magnetic field protection. Technologies such as protective shielding, real-time health monitoring, and exercise regimens aboard spacecraft are employed to mitigate these effects.

Despite the challenges, human space exploration aims to establish a permanent presence beyond Earth. NASA has announced plans to build a permanent base on the Moon by 2024, which will serve as a stepping stone for future missions to Mars. The colonization of space, also called space settlement or space humanization, envisions autonomous human habitats on other planets and moons, utilizing resources from those environments. This vision aligns with international agreements like the Outer Space Treaty, which prohibits national claims on celestial bodies and seeks to ensure space is used for the benefit of all humanity.

The issue of equitable participation and representation in space exploration has become increasingly important. International space law has long declared space "the province of all mankind," yet the benefits and opportunities of space exploration have historically been dominated by a few spacefaring nations. Organizations such as the JustSpace Alliance and initiatives like Inclusive Astronomy aim to promote greater diversity and inclusion in space exploration, ensuring that the rights of all nations and peoples are considered as humanity expands into space.

Women, in particular, have faced challenges in gaining equal representation in space programs. The first woman in space, Valentina Tereshkova, flew in 1963, but it wasn't until two decades later that another woman, Svetlana Savitskaya, followed in her footsteps. The United States sent its first female astronaut, Sally Ride, into space in 1983. While eleven countries have since sent women into space, female astronauts remain underrepresented, especially in long-duration missions. Issues such as the lack of appropriately sized spacesuits and concerns about health risks from radiation have limited women's participation in some space missions. Efforts are being made to address these challenges, and the first woman is expected to land on the Moon during the Artemis program in the near future.

The commercialization of space, coupled with advances in human spaceflight and astrobiology, continues to shape the future of space exploration. As private companies push the boundaries of what is possible in space tourism and commercial ventures, humanity's presence in space becomes more permanent and accessible. The ongoing search for life beyond Earth and the challenges of human space habitation drive the development of new technologies and international cooperation.

Chapter 2 - the Cold War

The Cold War, spanning from 1947 to 1991, was a period of intense geopolitical rivalry between the United States and the Soviet Union. This prolonged conflict, although lacking direct large-scale military engagement between the two superpowers, deeply influenced global politics, economics, and technology. It unfolded in the aftermath of World War II, when the United States and the Soviet Union, former allies in the fight against Nazi Germany, became locked in an ideological struggle that would shape international relations for nearly half a century.

The term "Cold War" refers to the absence of direct warfare between the superpowers. Instead, their rivalry manifested through proxy wars, where each supported opposing factions in regional conflicts, and in competition across various domains, including nuclear arms, economic influence, and space exploration.

The Cold War, a geopolitical and ideological struggle between the United States and the Soviet Union, emerged from the collapse of their wartime alliance after World War II. Between 1945 and 1949, the rivalry solidified, as both superpowers sought to assert influence over Europe and the world, leading to nearly half a century of tension between the Western Bloc, led by the United States, and the Eastern Bloc, dominated by the Soviet Union.

The roots of this conflict extended back to the early 20th century, with the 1917 Russian Revolution and the subsequent Treaty of Brest-Litovsk. This treaty, in which Soviet Russia ceded significant territory to Germany, raised concerns among the Western Allies. Further complicating relations, the Allied intervention in the Russian Civil War deepened Soviet distrust of the West. Although the Soviet Union later allied with the United States, Britain, and other Western powers during World War II to defeat Nazi Germany, this cooperation was fraught with suspicion. Each side viewed the other's actions through the lens of ideological rivalry, with the Soviets wary of capitalist expansionism, and the West distrustful of Soviet communism.

After World War II, the future of Eastern Europe became a major point of contention. The Soviet Union, having liberated much of Eastern Europe from Nazi control, established communist governments in these regions, bolstered by the presence of the Red Army. To Western leaders, this represented Soviet expansionism and a threat to democracy in Europe. In contrast, the Soviet leadership argued that they were ensuring their security by fostering governments aligned with Moscow.

In 1947, the introduction of the Marshall Plan, a U.S.-backed initiative to aid Europe's post-war reconstruction, further sharpened the divide. The Soviets rejected the plan, viewing it as a tool for the United States to impose its economic influence on Europe. In response, they established the Council for Mutual Economic Assistance (Comecon), aiming to strengthen economic cooperation among communist states.

The first significant confrontation of the Cold War unfolded with the Berlin Blockade in 1948. The Soviets sought to isolate West Berlin, cutting off road and rail access to the city. In response, the United States and its Western allies launched the Berlin Airlift, supplying West Berlin by air for nearly a year. This event marked a turning point, bringing the Cold War closer to direct military conflict. By 1949, Europe was firmly divided, with the creation of NATO solidifying military alliances in the West, while the Eastern Bloc fell under Soviet influence.

Tensions escalated outside of Europe as well, notably in Southeast Asia, where the Cold War spilled into proxy wars, such as the Korean War (1950-1953) and later the Vietnam War (1955-1975), as the U.S. sought to contain the spread of communism.

In early 1946, U.S. diplomat George F. Kennan sent his "Long Telegram" from Moscow, articulating the basis for the U.S. policy of containment toward the Soviet Union. This doctrine would guide American foreign policy for much of the Cold War. Around the same time, British Prime Minister Winston Churchill delivered his famous "Iron Curtain" speech, warning of Soviet expansionism and advocating for a firm Western response. Stalin's rebuttal was swift and sharp, accusing Churchill of advocating for Anglo-American dominance and comparing his rhetoric to Adolf Hitler's. The Soviet leader dismissed concerns about Soviet influence in Eastern Europe, arguing that Moscow was merely ensuring its security by establishing friendly governments.

The geopolitical struggle extended to other regions as well. The Soviet Union made territorial demands on Turkey and pressured Iran to maintain its military presence in the north, leading to a crisis. American pressure ultimately forced the Soviets to withdraw from Iran in 1946, marking an early success for the U.S. policy of containment.

By 1947, U.S. President Harry S. Truman's administration grew increasingly alarmed by Soviet actions in Eastern Europe and beyond. The Truman Doctrine, articulated in a speech to Congress, pledged U.S. support to countries resisting communist subjugation, particularly Greece and Turkey. This doctrine framed the Cold War as a global struggle between free democracies and totalitarian regimes, marking the beginning of a bipartisan consensus in the U.S. to counter Soviet influence worldwide. The doctrine was accompanied by financial aid, military assistance, and the establishment of alliances to contain Soviet expansion.

This consensus persisted for much of the Cold War, shaping U.S. and Western European foreign policy, even as dissenting voices emerged, particularly during and after the Vietnam War. Opposition came from various quarters, including anti-war activists and the nuclear disarmament movement, but the broader strategy of containment and deterrence remained central to Western efforts to confront the Soviet Union throughout the Cold War.

Thus, the early years of the Cold War set the stage for a protracted global confrontation that would span decades, affecting nearly every corner of the world as the two superpowers vied for ideological, political, and military supremacy.

In early 1947, the United States, France, and Britain sought to negotiate with the Soviet Union to rebuild a self-sufficient Germany, presenting a detailed plan that accounted for the industrial plants and infrastructure seized by the Soviets. However, these efforts failed, and tensions rose between the Western powers and the Soviet Union. In June 1947, in line with the Truman Doctrine, the United States introduced the Marshall Plan, a massive economic assistance program to rebuild war-torn Europe. The plan, signed by President Harry S. Truman on April 3, 1948, offered over $13 billion (equivalent to approximately $189 billion today) in aid to Western European countries. Its goals were to stabilize economies, prevent the spread of communism, and promote democratic governance across Europe.

The Marshall Plan was instrumental in revitalizing Europe's democratic and economic systems. However, it also highlighted the growing ideological divide between East and West. Stalin saw the plan as an American attempt to realign Europe toward capitalist interests and increase U.S. influence. Consequently, the Soviet Union rejected the Marshall Plan and prevented Eastern Bloc countries from accepting aid, instead establishing the Molotov Plan, which promoted economic cooperation among communist states. This division solidified as Eastern Europe fell deeper under Soviet control, and Stalin, wary of a resurgent Germany, ensured that the Soviet sphere of influence would not be threatened.

In early 1948, the Cold War escalated further with the Czechoslovak coup d'état. Communist forces, backed by the Soviet Union, seized control of Czechoslovakia, the last remaining democratic country in Eastern Europe. This event shocked the West, heightening fears of Soviet expansionism. The brutality of the coup eradicated any lingering opposition in the U.S. Congress to the Marshall Plan, ensuring its full implementation. The coup also marked a turning point, pushing the Cold War into a new phase of open hostility between the superpowers.

Following the coup, the London Six-Power Conference convened, ultimately leading to the Soviet boycott of the Allied Control Council, responsible for administering post-war Germany. This marked the end of hopes for a unified German state. By 1949, Germany was officially divided into two separate states: the Federal Republic of Germany (West Germany) and the German Democratic Republic (East Germany), cementing the division of Europe and the beginning of the full-scale Cold War.

Between 1948 and 1962, Cold War tensions escalated globally. The twin policies of the Truman Doctrine and the Marshall Plan provided billions of dollars in economic and military aid to Western Europe, Greece, and Turkey, aiding in their recovery and defense against communist influence. In Greece, U.S. support helped the government win a civil war against communist insurgents. In Italy, the Christian Democrats, under the leadership of Alcide De Gasperi, defeated the Communist-Socialist alliance in the 1948 elections, further stabilizing Western Europe against the spread of communism.

Espionage played a critical role throughout the Cold War, with both superpowers heavily relying on intelligence gathering. The Soviet Union's KGB, renowned for its effectiveness, spearheaded foreign espionage and internal surveillance. One of the KGB's most significant achievements was infiltrating the United States' Manhattan Project, which allowed the Soviets to acquire key information on nuclear technology. As a result, the Soviet Union detonated its first nuclear weapon in 1949, only four years after the United States' first test, much earlier than Western experts had anticipated.

On the Western side, the U.S. intelligence community also developed advanced espionage methods, particularly in signals intelligence. The Venona project, a top-secret U.S. effort to decrypt Soviet communications, provided substantial evidence of Soviet spy networks operating within America. Despite the project's successes, it remained highly classified, and even President Truman may not have been fully informed of its findings. The Venona project revealed the extent of Soviet infiltration, including the activities of prominent atomic spies who had helped accelerate the Soviet Union's nuclear program.

As the Cold War progressed, both the CIA and the KGB engaged in increasingly sophisticated espionage tactics. Soviet efforts included "active measures"—operations aimed at spreading disinformation, forging documents, and manipulating foreign media to sway international opinion in favor of the USSR. These clandestine activities were central to Soviet foreign policy, with KGB Major General Oleg Kalugin later describing them as the "heart and soul" of Soviet intelligence operations.

The espionage wars extended beyond the U.S.-Soviet rivalry. During the Sino-Soviet split in the 1960s, the USSR and the People's Republic of China engaged in their own intelligence battles, further complicating the global landscape of the Cold War.

The period from 1948 to 1962 saw the Cold War transform from a diplomatic standoff into a full-blown global confrontation, with Europe divided into two ideologically opposed blocs. As espionage and military tensions escalated, the world teetered on the brink of direct conflict between the superpowers, setting the stage for several of the most dangerous moments of the Cold War.

In September 1947, the Soviet Union established the Communist Information Bureau, or Cominform, to coordinate communist parties across the Eastern Bloc. This organization sought to tighten political control over Soviet satellite states and ensure ideological conformity within the international communist movement. However, the Cominform faced a significant setback in June 1948, when the Tito–Stalin split forced Yugoslavia's expulsion from the organization. Josip Broz Tito, Yugoslavia's leader, resisted Soviet interference in his country's internal affairs, adopting a non-aligned position that placed him at odds with Stalin. Despite remaining a communist state, Yugoslavia began to accept financial aid from the United States, signaling its independence from Soviet influence.

The split between Tito and Stalin also affected broader Cold War geopolitics, including the contested status of the city of Trieste, a flashpoint of Cold War tensions. The Free Territory of Trieste, established in 1947 under the United Nations, served as a neutral buffer zone between Italy and Yugoslavia, reflecting the complex dynamics of the Cold War. The area was divided, and control was disputed by various factions, including Italians, Slovenes, monarchists, and republicans. However, with the détente between Tito's Yugoslavia and the West, the territory was dissolved in 1954 and 1975, ending the disputes over Trieste.

At the same time, the Berlin Blockade became one of the earliest and most significant crises of the Cold War. The US and Britain had merged their occupation zones in western Germany into the "Bizone" in January 1947, with France joining in April 1949 to form the "Trizone." In early 1948, as part of the Marshall Plan, Western leaders initiated efforts to rebuild the German economy, including introducing a new currency, the Deutsche Mark, which replaced the devalued Reichsmark. This economic revitalization angered the Soviets, who viewed the West's efforts as a direct challenge to their control over East Germany.

In response, in June 1948, Stalin imposed the Berlin Blockade, cutting off all ground access to West Berlin, an enclave within Soviet-controlled East Germany. Stalin aimed to force the Western Allies to abandon West Berlin or agree to Soviet demands. Instead, the US, Britain, France, and their allies launched the Berlin Airlift, a massive operation to supply West Berlin with food, fuel, and other necessities. Planes from the US, Britain, and other countries, including Canada, Australia, and New Zealand, flew thousands of missions, delivering essential goods to the besieged city despite Soviet threats and efforts to disrupt the operation.

The airlift became a powerful symbol of Western resolve, and public support grew in West Berlin and abroad. In December 1948, municipal elections in Berlin saw a resounding victory for non-communist parties, with 86.3% voter turnout. The results solidified the city's political division into East and West Berlin. In a display of solidarity, 300,000 West Berliners rallied to support the airlift, which became known as "Operation Vittles." One of the most enduring images of the operation was US Air Force pilot Gail Halvorsen's initiative, "Operation Little Vittles," where he dropped candy to the children of Berlin, winning hearts and minds during this tense period.

The Berlin Airlift was a logistical and political success for the Western Allies, reinforcing the link between West Berlin and the United States. By May 1949, Stalin conceded defeat and lifted the blockade, allowing normal ground traffic to resume. The crisis cemented Berlin's division and deepened the Cold War divide between East and West.

Despite the failure of the blockade, Stalin continued to propose plans for a unified Germany. In 1952, he repeatedly suggested that East and West Germany could be unified under a single government elected through United Nations-supervised elections on the condition that the new Germany remain neutral and refrain from joining Western military alliances. However, the Western powers rejected the proposal, questioning its sincerity and suspecting that Stalin's true aim was to neutralize West Germany and extend Soviet influence. These events marked the beginning of the protracted Cold War rivalry that would define European and global geopolitics for decades to come.

In April 1949, Britain, France, the United States, Canada, and eight other Western European countries signed the North Atlantic Treaty, officially establishing the North Atlantic Treaty Organization (NATO). This defensive military alliance aimed to provide collective security against the growing threat posed by the Soviet Union and its expanding influence in Eastern Europe. NATO's formation marked a pivotal moment in the Cold War, as it formally united the Western powers against potential Soviet aggression.

Later that year, in August 1949, the Soviet Union successfully tested its first atomic bomb in Semipalatinsk, Kazakh SSR, dramatically altering the global balance of power. The United States was no longer the sole nuclear power, and the arms race between the two superpowers had begun. Meanwhile, the U.S., Britain, and France took another significant step in Western Europe by establishing the Federal Republic of Germany (West Germany) in May 1949, composed of the three Western occupation zones. This action came after the Soviet Union had refused to participate in a joint German rebuilding effort proposed by the Western European countries in 1948. In response, the Soviet Union declared its occupation zone the German Democratic Republic (East Germany) in October 1949, further entrenching the division of Europe into East and West.

The media landscape in the Eastern Bloc reflected the authoritarian nature of the Soviet-controlled regimes. All radio and television networks were state-owned, and communist parties usually controlled the print media. These outlets served as tools for disseminating Soviet propaganda, promoting Marxist ideology while attacking capitalism.

Soviet broadcasts emphasized themes of labor exploitation, imperialism, and the perceived warmongering of the West.

To counter Soviet influence, the United States and its allies launched their own propaganda initiatives. Radio Free Europe, established in 1949, was a key part of this effort. Broadcasting to Central and Eastern Europe provided an alternative to the censored and state-controlled media of the Eastern Bloc. Alongside the BBC and Voice of America, Radio Free Europe sought to undermine communist regimes by offering uncensored news and commentary, presenting a different narrative to the official Soviet line. The broadcasts were a product of Cold War strategists like George F. Kennan, who believed that the Cold War would ultimately be won through political rather than military, means. Soviet and Eastern Bloc authorities, in turn, attempted to suppress these Western broadcasts through radio jamming and other forms of interference.

The ideological nature of the Cold War was not lost on American policymakers. Figures like Kennan and John Foster Dulles recognized that the conflict was, at its core, a battle of ideas. In response, the United States, through the Central Intelligence Agency (CIA), funded a range of initiatives designed to counter communist influence, particularly among intellectuals in Europe and the developing world. One such project was the "Crusade for Freedom," a domestic propaganda campaign to rally public support for Cold War objectives.

The question of German rearmament became a contentious issue in the early 1950s. West Germany's chancellor, Konrad Adenauer, strongly advocated rearmament, viewing it as essential for his country's sovereignty and defense within the Western alliance. France, however, was initially opposed, fearing the resurgence of German militarism. The United States played the decisive role in the debate, with the Pentagon advocating for West Germany's rearmament, while President Truman and the State Department remained more ambivalent. The outbreak of the Korean War in June 1950 changed the calculus, as the U.S. now saw a rearmed West Germany as essential to the defense of Europe. General Dwight D. Eisenhower was appointed Supreme Commander of NATO forces, and additional U.S. troops were sent to West Germany.

Rearming West Germany culminated in its full NATO membership in 1955. This development came with assurances that West Germany would not develop nuclear weapons, addressing widespread concerns about the potential return of German militarism. West Germany's rearmament and integration into NATO were framed within a broader alliance structure, ensuring its military forces would operate under NATO command.

Not everyone supported the Western alignment of Germany. In 1953, Lavrentiy Beria, a prominent Soviet leader, proposed reunifying Germany as a neutral state to prevent its incorporation into NATO. However, Beria's proposal failed to gain traction, and his execution later that year during a power struggle in the Soviet Union ended any further attempts to pursue this strategy. By 1955, the Bundeswehr, West Germany's new military, was formally established, marking the full integration of West Germany into the Western defense system and further solidifying the division of Europe.

In 1949, the culmination of the Chinese Civil War marked a pivotal shift in global geopolitics. Mao Zedong's People's Liberation Army defeated the United States-backed Kuomintang (KMT) Nationalist Government, forcing its leader, Chiang Kai-shek, to retreat to Taiwan. This established the People's Republic of China (PRC) on the mainland. The Soviet Union, under Joseph Stalin, quickly allied with Mao's new communist regime. Historian Odd Arne Westad attributes the Communist victory to strategic advantages—Mao's forces made fewer military mistakes and garnered broader support by appealing to various interest groups, including peasants, while Chiang's efforts to centralize power alienated many within China. Chiang's KMT, already weakened by the war against Japan, struggled to maintain cohesion, while the Communists skillfully cloaked their revolutionary agenda in Chinese nationalism.

This power shift, coupled with the Soviet Union's successful development of atomic weapons in 1949, significantly altered the strategic landscape. The Truman administration responded by escalating its policy of containment, outlined in the top-secret National Security Council document NSC 68. This policy called for a dramatic increase in U.S. defense spending and the reinforcement of alliances in Europe and Asia, Africa, and Latin America. Paul Nitze, a key advisor, influenced President Truman to view containment as a comprehensive strategy to roll back Soviet influence worldwide.

In the early 1950s, the U.S. expanded its network of alliances, a strategy sometimes referred to as "Pactomania." These alliances included key agreements such as ANZUS in 1951 and the Southeast Asia Treaty Organization (SEATO) in 1954. Through these pacts, the U.S. secured military bases across strategic regions, including Japan, South Korea, Taiwan, Australia, New Zealand, Thailand, and the Philippines, solidifying its Pacific and East Asia presence.

The Korean War began in June 1950 and served as a critical test of this expanded containment policy. After years of rising tensions, North Korean forces under Kim Il Sung crossed the 38th parallel, invading South Korea. Though initially reluctant, Stalin supported the invasion by sending military advisors. With the United States leading a coalition of 16 nations, the United Nations responded swiftly, passing Resolutions 82 and 83 to defend South Korea. Despite the Soviet Union's boycott of the Security Council over Taiwan's representation, the UN approved military intervention. The initial U.S.-led effort aimed to repel North Korean forces and restore the boundary at the 38th parallel. However, after the successful Inchon landing, the U.S. adopted a more ambitious rollback strategy, crossing into North Korea to unify the peninsula under a democratic government. This shift provoked China, which feared U.S. expansion into its sphere of influence, to intervene with a massive military force, pushing UN troops back into South Korea.

The Korean War ultimately settled into a stalemate near the original border, solidifying the division of the Korean Peninsula. The conflict had far-reaching consequences, galvanizing the creation of a more robust NATO military structure and reinforcing the wisdom of the containment doctrine over the riskier rollback strategy.

Following the war, the political landscape in both Koreas evolved in starkly different directions. In the North, Kim Il Sung established a centralized, totalitarian regime that promoted a pervasive cult of personality. In the South, U.S.-backed dictator Syngman Rhee ruled through authoritarian means, suppressing communist sympathies with brutal force. Although Rhee was deposed in 1960, South Korea continued under military rule until the late 1980s, when it transitioned to a democratic system and experienced rapid economic growth, becoming a global economic powerhouse.

The death of Joseph Stalin in 1953 and the election of Dwight D. Eisenhower as U.S. president that same year marked a shift in the dynamics of the Cold War. In seeking to balance Cold War objectives with domestic concerns, Eisenhower reduced military spending by a third while maintaining a strong stance against Soviet aggression. In the Soviet Union, a power struggle ensued, with Nikita Khrushchev eventually emerging as the leader. Khrushchev's de-Stalinization policies, including his famous denunciation of Stalin in 1956, aimed to soften Soviet control over both society and the Communist Party. His rhetoric, however, remained aggressive. In 1956, Khrushchev famously declared, "We will bury you," alarming Western diplomats. He later clarified that he referred to communism's eventual victory over capitalism, not imminent war.

Bold promises of Soviet progress marked Khrushchev's tenure. He predicted that within a decade, the Soviet Union's housing crisis would be resolved, consumer goods would be plentiful, and the construction of a communist society would be well underway. These declarations of Soviet superiority became a central theme in the Cold War rivalry.

Eisenhower's administration, particularly under Secretary of State John Foster Dulles, responded with a "New Look" strategy that emphasized reliance on nuclear weapons to deter Soviet aggression. The doctrine of "massive retaliation" threatened to overwhelm the U.S. nuclear response to any Soviet or communist provocation. This approach proved effective during the Suez Crisis of 1956, where the U.S.'s nuclear capabilities neutralized Soviet threats to intervene.

Despite these tensions, there were moments of thaw in Cold War diplomacy. In 1959, Khrushchev made an unprecedented visit to the United States, signaling a potential shift toward détente. However, hopes for lasting peace were dashed by the U-2 spy plane incident in 1960, where an American reconnaissance aircraft was shot down over Soviet territory. Eisenhower's subsequent denial of the mission, followed by the Soviet Union's exposure of the cover-up, reignited Cold War hostilities and undermined the prospects for a two-power summit.

In the aftermath of Joseph Stalin's death in 1953, tensions within Europe experienced a slight easing, yet the continent remained entrenched in an uneasy armed truce. The Soviet Union established a network of mutual assistance treaties throughout the Eastern Bloc by 1949 and formalized this alliance by creating the Warsaw Pact in 1955. This military alliance directly opposed the North Atlantic Treaty Organization (NATO), solidifying the division of Europe into competing spheres of influence.

The Warsaw Pact symbolized Soviet dominance over Eastern Europe and served as a strategic counterbalance to NATO's collective defense mechanism. This period marked the zenith of Soviet territorial influence, which would later recede following significant geopolitical shifts such as the Cuban Revolution of 1959 and the Sino-Soviet split of 1961.

A pivotal moment in Cold War history occurred in Hungary in 1956, illustrating the fragility of Soviet control over its Eastern European allies. The Hungarian Revolution of October 1956 erupted shortly after Soviet Premier Nikita Khrushchev orchestrated the removal of Hungary's staunch Stalinist leader, Mátyás Rákosi. Inspired by widespread anti-communist sentiment, Hungarian protesters demanded the dissolution of the secret police, withdrawal from the Warsaw Pact, and the establishment of free elections. In response, the Soviet Army launched a brutal invasion to suppress the uprising. The conflict resulted in the deaths of thousands of Hungarians, the imprisonment and deportation of countless others, and the exodus of approximately 200,000 refugees fleeing the chaos. Hungarian leader Imre Nagy and several revolutionaries were executed following secret trials, underscoring the Soviet Union's unwavering commitment to maintaining its grip on Eastern Europe.

The suppression of the Hungarian Revolution had profound repercussions beyond Hungary's borders. It exposed deep ideological fractures within global communist movements, particularly within Western Europe. The violent response to the uprising led to a significant decline in membership and support for communist parties in the West, as disillusionment grew among both Western and socialist nations. Prominent figures such as Yugoslavian politician Milovan Đilas remarked that "the wound which the Hungarian Revolution inflicted on communism can never be completely healed," highlighting the event's lasting impact on the international communist movement.

Amid these developments, the Soviet Union continued to assert its military prowess. Between 1957 and 1961, Khrushchev frequently threatened the West with nuclear annihilation, boasting that Soviet missile capabilities surpassed those of the United States. He claimed that the Soviet Union possessed the means to obliterate any American or European city, thereby heightening global fears of nuclear conflict. However, historian John Lewis Gaddis notes that Khrushchev departed from Stalin's "belief in the inevitability of war," instead advocating for "peaceful coexistence." In Khrushchev's vision, peaceful coexistence would allow capitalism to collapse organically while providing the Soviet Union time to enhance its military capabilities—a strategy that persisted until Mikhail Gorbachev's later policies redefined the concept of peaceful coexistence.

The Hungarian Revolution also influenced broader Cold War dynamics, particularly the nuclear disarmament discourse. In 1957, Polish Foreign Minister Adam Rapacki proposed the Rapacki Plan, advocating for a nuclear-free zone in Central Europe. While the proposal garnered favorable public opinion in the West, it was ultimately rejected by leaders of West Germany, Britain, France, and the United States. These nations feared that the absence of nuclear weapons would leave the Warsaw Pact's conventional armies unchallenged, thereby destabilizing the existing balance of power.

Concurrently, tensions over Berlin continued to escalate. In November 1958, Khrushchev attempted to transform Berlin into an independent, demilitarized "free city." He issued a six-month ultimatum to the United States, Great Britain, and France, demanding the withdrawal of their troops from West Berlin or threatening to transfer control of Western access rights to East Germany. Khrushchev infamously declared to Chinese Premier Mao Zedong, "Berlin is the testicles of the West. Whenever I want to make the West scream, I squeeze on Berlin." The ultimatum was swiftly rejected by NATO in December 1958, leading Khrushchev to withdraw his demands in exchange for a Geneva conference to address the German question.

As the late 1950s progressed, the United States recognized the need to adapt its Cold War strategy in response to evolving Soviet tactics. Under President Dwight D. Eisenhower, the U.S. had previously implemented the "New Look" policy, which emphasized the use of nuclear weapons to deter Soviet aggression while reducing expenditures on conventional forces. This approach sought to leverage the cost-effectiveness of nuclear deterrence, allowing the United States to maintain a formidable defense posture without the financial burden of a large standing army.

However, John F. Kennedy's election in 1960 signaled a strategic shift. Kennedy's foreign policy was characterized by heightened confrontations with the Soviet Union, often manifested through proxy conflicts around the globe. Building on the foundation of containment established by his predecessors, Kennedy introduced the doctrine of "flexible response." This strategy aimed to provide the United States with a range of military options, from conventional forces to nuclear capabilities, enabling limited and proportional responses to Soviet provocations without immediately resorting to nuclear warfare.

Kennedy initiated a substantial increase in defense spending to support the flexible response strategy. He advocated for the rapid expansion of the nuclear arsenal to reclaim perceived losses in Soviet superiority, criticizing Eisenhower for prioritizing budget deficits over military strength. In his inaugural address, Kennedy pledged "to bear any burden" to defend liberty, emphasizing the necessity of enhanced military capabilities. Between 1961 and 1964, the United States saw a 50 percent increase in nuclear weapons, a similar rise in B-52 bombers, and a significant expansion of intercontinental ballistic missiles (ICBMs) from 63 to 424 units. Additionally, Kennedy authorized the deployment of 23 new Polaris submarines, each equipped with 16 nuclear missiles, and urged American cities to construct fallout shelters in preparation for potential nuclear conflict.

Kennedy's administration also expanded the United States' special operations forces, elite military units trained for unconventional warfare, further diversifying the nation's strategic options. This expansion aimed to enable the United States to counter Soviet influence and address global threats through versatile military engagements, thereby reducing the reliance on nuclear deterrence alone.

The period from the mid-1950s to the early 1960s was marked by intense geopolitical maneuvering as both the Soviet Union and the United States sought to assert their dominance on the global stage. Events such as the Hungarian Revolution, the Rapacki Plan, and the Berlin Crisis underscored the volatile nature of Cold War alliances and the ever-present threat of nuclear confrontation. Concurrently, the United States' strategic military buildup under Kennedy set the stage for an era of heightened competition that would soon extend into the realm of space exploration, igniting the Space Race as the latest frontier in the struggle for global supremacy.x

In the aftermath of World War II, the collapse of European colonial empires in Asia and Africa set the stage for a fierce competition between the United States and the Soviet Union for influence in the newly independent nations of the Third World. Nationalist movements in countries such as Guatemala, Indonesia, and Indochina often found themselves allied with communist factions or perceived as adversarial to Western interests. As decolonization gained momentum throughout the 1950s and early 1960s, both superpowers sought to extend their ideological reach by supporting allied governments and undermining those deemed hostile or neutral.

The United States, wary of the spread of communism, increasingly relied on the Central Intelligence Agency (CIA) to execute covert operations aimed at maintaining and expanding its influence. In 1953, under President Dwight D. Eisenhower, the CIA orchestrated Operation Ajax, a clandestine coup to overthrow Iran's democratically elected Prime Minister Mohammad Mosaddegh. Mosaddegh had nationalized the British-owned Anglo-Iranian Oil Company in 1951, inciting significant tensions with Britain and the United States. Faced with accusations from British Prime Minister Winston Churchill that Mosaddegh was succumbing to communist influence, the Eisenhower administration supported the return of the pro-Western Shah, Mohammad Reza Pahlavi, to power. The Shah's regime subsequently banned the communist Tudeh Party of Iran and established the SAVAK, a brutal security and intelligence agency tasked with suppressing political dissent.

Similarly, the United States intervened decisively in Guatemala in 1954 in Central America. The democratically elected President Jacobo Árbenz had implemented progressive land reforms that threatened the interests of the United Fruit Company, an American corporation with extensive holdings in the region. Perceived as left-leaning and potentially susceptible to Soviet influence, Árbenz was ousted in a CIA-supported coup. The ensuing military junta, led by Carlos Castillo Armas, reversed Árbenz's reforms, restored properties to the United Fruit Company, and established anti-communist institutions, including the National Committee of Defense Against Communism and the Preventive Penal Law Against Communism. These actions effectively dismantled Guatemala's progressive agenda and reinforced U.S. dominance in the region.

In Southeast Asia, the struggle for independence and the subsequent Cold War dynamics led to significant turmoil in Indonesia. The non-aligned government of President Sukarno faced internal threats from regional commanders demanding autonomy. By 1958, these dissident military leaders, including Colonel Ahmad Husein in Central Sumatra and Colonel Ventje Sumual in North Sulawesi, formed the Permesta Movement to overthrow Sukarno's regime. Supported by anti-communist elements, the rebels received covert assistance from the CIA. The conflict intensified when American pilot Allen Lawrence Pope was shot down during a bombing raid on government-held Ambon in April 1958. In response, the central government launched comprehensive military operations, including airborne and seaborne invasions of rebel strongholds in Padang and Manado. By the end of 1958, the Permesta rebels were militarily defeated, and the last guerrilla bands surrendered by August 1961, consolidating Sukarno's authority.

The Republic of the Congo, newly independent from Belgium in June 1960, quickly descended into crisis. On July 5, 1960, the Congo Crisis erupted, leading to the secession of the mineral-rich regions of Katanga and South Kasai. President Joseph Kasa-Vubu, backed by the CIA, dismissed the democratically elected Prime Minister Patrice Lumumba in September 1960, citing Lumumba's alleged ties to the Soviet Union and his role in the massacres during the invasion of South Kasai. Colonel Mobutu Sese Seko, with CIA support, seized power through a military coup, ultimately imprisoning Lumumba and handing him over to Katangan authorities, who executed him by firing squad. The Congo Crisis underscored the United States' commitment to preventing Soviet influence in Africa, often at the expense of democratic governance and stability.

In British Guiana (now Guyana), the leftist People's Progressive Party (PPP) emerged as a significant political force in the 1950s. Cheddi Jagan, the PPP's candidate, won the position of chief minister in the 1953 colonial elections. However, Britain suspended the nation's constitution and subsequent intervention forced Jagan to resign. Despite Jagan's repeated electoral victories in 1957 and 1961, the British, influenced by U.S. pressure, continued to undermine his administration by imprisoning PPP leaders and fostering divisions within the party. This manipulation delayed Guyana's independence and ensured that a more pro-Western government would assume power, aligning the nation with U.S. interests.

The decolonization process in Vietnam exemplified the intense Cold War rivalry in the Third World. After enduring a protracted guerrilla war against French colonial forces, the Viet Minh, led by Ho Chi Minh, achieved a decisive victory at the Battle of Dien Bien Phu in 1954. This defeat compelled France to withdraw from Vietnam, leading to the Geneva Conference, where peace accords were signed. The accords divided Vietnam at the 17th parallel, establishing a pro-Soviet administration in the North and a pro-Western government in the South. From 1954 to 1961, the United States provided substantial economic aid and military advisers to bolster South Vietnam's defenses against communist insurgencies, laying the groundwork for future conflict in the region.

Amid these proxy battles, many newly independent nations sought to navigate a path independent of the bipolar Cold War dynamics. The Bandung Conference of 1955, held in Indonesia, brought together representatives from dozens of Third World countries to promote solidarity and cooperation while rejecting alignment with either the United States or the Soviet Union. This movement towards non-alignment culminated in establishing the Non-Aligned Movement in 1961, headquartered in Belgrade. Leaders such as India's Jawaharlal Nehru and Yugoslavia's Josip Broz Tito sought to create a third path emphasizing national sovereignty and mutual cooperation without succumbing to superpower pressures.

During this era, the Soviet Union also sought to expand its influence by forging ties with key neutral states. Nikita Khrushchev's administration aimed to build relationships with countries like India, which played a pivotal role in the Non-Aligned Movement. These efforts were part of a broader strategy to present the Soviet Union as a supporter of anti-colonial and nationalist movements, thereby enhancing its global standing and countering Western influence without direct confrontation.

A complex interplay of ideology, economic interests, and geopolitical strategy characterized the competition in the Third World. As decolonization transformed the global order, the United States and the Soviet Union engaged in a multifaceted struggle for supremacy, utilizing a combination of military interventions, covert operations, economic aid, and diplomatic alliances. This intense rivalry not only shaped the political landscapes of emerging nations but also set the stage for the subsequent Space Race, as both superpowers sought to demonstrate their technological and ideological superiority on the global stage.

In the early 1960s, the Cold War reached new heights of tension and confrontation, profoundly influencing global politics and the ongoing Space Race between the United States and the Soviet Union. Central to this era were pivotal events in Cuba and Berlin, which shaped international relations and underscored the pervasive fear of nuclear conflict that loomed over technological competitions in space.

On January 1, 1959, the 26th of July Movement, led by the charismatic Fidel Castro and the idealistic Ernesto "Che" Guevara, successfully overthrew the regime of Cuban President Fulgencio Batista. Batista's dictatorship had long been marred by corruption and repression, and his unpopularity was further exacerbated by the Eisenhower administration's refusal to supply arms to his government. Initially, Castro resisted labeling his new government as socialist, frequently denying any association with communism. However, his administration swiftly appointed Marxists to key positions, signaling a clear shift toward socialist policies. Che Guevara, for instance, was appointed Governor of the Central Bank before ascending to the role of Minister of Industries, thereby embedding Marxist ideology within the Cuban government's framework.

In the aftermath of Batista's ousting, diplomatic relations between Cuba and the United States remained formally intact but were fraught with underlying tensions. During Castro's visit to Washington, D.C., in April 1960, President Dwight D. Eisenhower deliberately avoided meeting him, delegating the responsibility to Vice President Richard Nixon instead. This subtle snub highlighted the growing unease in U.S.-Cuban relations. By March 1960, Cuba had begun negotiating arms purchases from the Eastern Bloc, indicating a pivot toward Soviet support. Simultaneously, the Eisenhower administration sanctioned the CIA to develop plans and allocate funding aimed at overthrowing Castro's government, laying the groundwork for future covert operations.

The diplomatic rift deepened in January 1961 when President Eisenhower formally severed diplomatic relations with Cuba shortly before leaving office. This escalation set the stage for the Bay of Pigs Invasion in April 1961.

Under the new administration of President John F. Kennedy, the United States orchestrated an unsuccessful CIA-backed amphibious invasion by Cuban exiles at Playa Girón and Playa Larga in Santa Clara Province. The failure of the invasion not only publicly humiliated the United States but also emboldened Castro to embrace Marxism–Leninism openly. In response, the Soviet Union pledged increased support to Cuba, further aligning the island nation with the Eastern Bloc.

Concurrently, Berlin remained a focal point of Cold War tensions. By the early 1950s, the Soviet Union had implemented strict measures to restrict emigration from East Germany, a policy mirrored by other Eastern Bloc countries. Despite these efforts, hundreds of thousands of East Germans continued to flee to West Germany each year, exploiting a loophole between East and West Berlin. This mass migration resulted in a significant "brain drain," as a substantial portion of East Germany's young, educated population migrated to the prosperous West, undermining the East's economic and social stability.

In June 1961, the Soviet Union issued an ultimatum demanding the withdrawal of Allied forces from West Berlin, a request that the United States firmly rejected. Instead, the U.S. limited its security guarantees to West Berlin alone. On August 13, East Germany responded by erecting a barbed-wire barrier around West Berlin, which would soon be transformed into the formidable Berlin Wall. This construction effectively sealed the border, halting the exodus of East Germans and symbolizing the deepening division between the Eastern and Western blocs.

Amid these geopolitical upheavals, the Space Race continued to serve as a critical arena for demonstrating technological and ideological superiority. However, the crises in Cuba and Berlin underscored the ever-present threat of nuclear confrontation, which cast a shadow over the pursuit of space exploration milestones.

The culmination of these tensions occurred during the Cuban Missile Crisis of October 1962, a pivotal moment that brought the world perilously close to nuclear war. Following the failed Bay of Pigs Invasion, the Kennedy administration intensified efforts to destabilize Castro's government through Operation Mongoose—a covert campaign involving sabotage and terrorist activities aimed at overthrowing the Cuban regime. In response to these aggressive U.S. actions and to bolster Cuba's defense, Soviet Premier Nikita Khrushchev authorized the installation of nuclear missiles on Cuban soil.

When American reconnaissance flights captured aerial photographs of Soviet missile sites in Cuba in November 1962, President Kennedy faced an unprecedented crisis. After careful deliberation, he opted to impose a naval blockade, termed a "quarantine," around Cuba to prevent further Soviet shipments of military equipment. Kennedy demanded the removal of the missiles and the dismantling of the existing sites. The world watched as Soviet and American tanks faced off at Checkpoint Charlie in Berlin, a stark symbol of the brinkmanship that defined the era.

Intense negotiations between Kennedy and Khrushchev ultimately averted nuclear war. The Soviet Union agreed to withdraw its missiles from Cuba in exchange for a public American pledge not to invade the island and a covert agreement to remove U.S. missiles from Turkey. However, the resolution of the crisis had significant political repercussions. Khrushchev's concession was perceived as a retreat, damaging his prestige and contributing to his eventual ousting in 1964. The Soviet leadership criticized him for mismanaging Soviet agriculture, nearly provoking nuclear war, and constructing the Berlin Wall, viewing these actions as detrimental to Soviet interests and global standing.

The Cuban Missile Crisis underscored the fragile balance of power during the Cold War and highlighted the urgent need for nuclear arms control. In its aftermath, both superpowers recognized the necessity of improving diplomatic relations and preventing future escalations. While the Antarctic Treaty of 1961 had already marked a step toward international cooperation, the crisis emphasized the importance of continued efforts to mitigate the risks of nuclear confrontation.

These critical events in Cuba and Berlin not only shaped the geopolitical landscape of the early 1960s but also influenced the dynamics of the Space Race. As both the United States and the Soviet Union sought to outpace each other in space exploration, the underlying tensions and the ever-present threat of nuclear conflict remained a constant backdrop. This environment drove the relentless pursuit of technological superiority and strategic advantage, with each superpower striving to demonstrate its dominance not just on Earth, but in the vast expanse of space as well.

As the 1960s progressed into the late 1970s, the Cold War underwent significant transformations, marked by shifting alliances, economic upheavals, and a gradual movement from direct confrontation towards détente. This period was characterized by a more complex international landscape, where the binary division of the world into two opposing blocs became increasingly blurred.

In the aftermath of World War II, Western Europe and Japan experienced rapid recovery and robust economic growth, largely supported by American aid through initiatives such as the Marshall Plan. By the 1950s and 1960s, these regions enjoyed per capita GDPs approaching those of the United States. In stark contrast, the economies of the Eastern Bloc stagnated under the rigid control of Soviet-style centralized planning. This economic disparity underscored the inherent weaknesses of the communist system, fueling dissatisfaction and unrest within the Soviet sphere.

Amidst this backdrop, the United States found itself entangled in the protracted and draining conflict of the Vietnam War. Initially escalating under President John F. Kennedy, who increased US troop levels from fewer than a thousand in 1959 to 16,000 by 1963, the war became a quagmire that severely impacted American international prestige and economic stability. The heavy-handed tactics of South Vietnamese President Ngo Dinh Diem, particularly his crackdown on Buddhist monks in 1963, led the US to support a military coup, further destabilizing the region.

The situation deteriorated following the Gulf of Tonkin incident in 1964, where alleged clashes between US destroyers and North Vietnamese forces led to the Gulf of Tonkin Resolution. This gave President Lyndon B. Johnson broad authority to escalate US military involvement, resulting in the deployment of ground combat units and increased troop levels to 184,000. Soviet Premier Leonid Brezhnev responded by reversing Khrushchev's policy of disengagement, providing increased aid to North Vietnam to draw it away from its pro-Chinese stance. Despite this support, the Soviet Union limited its involvement to avoid further escalation, offering enough assistance to entangle American forces without provoking a broader conflict.

The Tet Offensive in 1968 marked a pivotal turning point in the Vietnam War. This massive and coordinated assault by North Vietnamese and Viet Cong forces demonstrated the vulnerability of South Vietnamese defenses and the limits of American military power. Although the offensive was ultimately repelled, it shattered the illusion of imminent victory and intensified domestic opposition to the war in the United States. The resulting "Vietnam Syndrome" reflected a widespread aversion to American military interventions abroad, influencing US foreign policy for years to come.

Simultaneously, economic and political shifts were reshaping the global geopolitical landscape. The 1973 oil crisis, triggered by the Organization of Petroleum Exporting Countries (OPEC) cutting its oil production, profoundly affected Western economies, leading to skyrocketing oil prices and economic strain. Conversely, the Soviet Union benefited from the surge in oil revenues, which temporarily alleviated some of its own economic difficulties. This crisis also empowered Third World nations and movements, such as the Non-Aligned Movement, granting them greater leverage and independence from the superpowers.

These economic challenges, coupled with the rising influence of emerging nations, compelled the Soviet leadership to turn inward and address persistent domestic economic issues. Leaders like Leonid Brezhnev and Alexei Kosygin began to advocate for détente—a strategic easing of tensions between the Soviet Union and the United States. This policy shift aimed to reduce the risk of nuclear confrontation and foster a more stable international environment.

Within NATO, unity was tested by France's partial withdrawal from military structures under President Charles de Gaulle. De Gaulle, dissatisfied with what he perceived as the disproportionate influence of the United States within the alliance, sought to assert French independence. In 1966, he withdrew France from NATO's integrated military command and expelled NATO troops from French territory. This move highlighted the complexities of maintaining a unified Western front amid divergent national interests and strategies.

The pursuit of détente manifested in several key agreements and summits to reduce tensions and foster cooperation. The Strategic Arms Limitation Talks (SALT), initiated in the late 1960s, sought to curb the arms race by limiting the number of nuclear weapons each superpower could maintain. These negotiations represented a significant shift from the adversarial stance of earlier decades, reflecting a mutual recognition of the catastrophic potential of nuclear war.

Economic factors continued to play a crucial role in shaping Cold War dynamics. The Soviet Union's reliance on oil exports made it vulnerable to fluctuations in global oil markets, while Western economies struggled with inflation and energy shortages. These economic pressures necessitated reevaluating military and foreign policies, further encouraging the move towards détente.

In the realm of space exploration, the easing of Cold War tensions had a notable impact on the Space Race. While competition between the United States and the Soviet Union remained fierce, there were increasing opportunities for collaboration and joint ventures in space. The Apollo-Soyuz Test Project of 1975, where American and Soviet spacecraft docked in orbit, symbolized a newfound willingness to work together despite broader geopolitical rivalries. This cooperative spirit paved the way for future international partnerships in space exploration, highlighting the potential for shared human achievements even amidst political tensions.

By the late 1970s, détente had established a more stable and predictable framework for international relations, although underlying tensions and rivalries persisted. The period from 1962 to 1979 witnessed a gradual shift from direct confrontation to strategic accommodation, influenced by economic challenges, shifting alliances, and a mutual understanding of the dangers posed by nuclear proliferation. This era laid the groundwork for subsequent developments in the Cold War, including the eventual easing of hostilities and the eventual thawing of East-West relations.

The transition from confrontation to détente not only reshaped global politics but also influenced the trajectory of the Space Race. As both superpowers sought to balance military competition with opportunities for cooperation, exploring space became a domain where collaboration could potentially transcend terrestrial conflicts. This period of détente thus played a pivotal role in shaping the future of space exploration, fostering an environment where scientific and technological advancements could flourish alongside efforts to maintain global peace and stability.

The Era of Stagnation and Shifting Policies

The period of détente was not without its challenges. The Soviet Union, under leaders like Leonid Brezhnev, grappled with internal economic stagnation and the need to sustain its superpower status amidst declining growth rates. The concept of the "Era of Stagnation" emerged as Soviet economic performance faltered, revealing the limitations of centralized planning and the inefficiencies inherent in the Soviet economic model. This stagnation forced the Soviet leadership to prioritize stability and incremental reforms over ambitious expansion, further reinforcing the need for détente to manage international relations more pragmatically.

In conclusion, the years from 1962 to 1979 were marked by a significant evolution in Cold War dynamics. The shift from direct confrontation to détente reflected broader changes in the global economy, the rise of influential Third World nations, and the pressing need to address internal economic challenges within the superpowers. These developments not only altered the course of international relations but also had profound implications for the Space Race, highlighting the intricate interplay between political strategies and technological competition in shaping the modern world.

During the Cold War, Finland occupied a unique and precarious position, officially maintaining a stance of neutrality while being geographically and politically sandwiched between the Western bloc and the Soviet Union. This delicate balance was formalized through the YYA Treaty (Finno-Soviet Pact of Friendship, Cooperation, and Mutual Assistance) signed in 1948. The treaty granted the Soviet Union significant influence over Finnish domestic policies, a phenomenon later termed "Finlandization" by Western observers.

Finlandization entailed a range of self-imposed restrictions within Finland to avoid antagonizing the Soviet Union. The Soviet Union's influence subtly shaped Finnish politics and media, leading to widespread self-censorship and promoting pro-Soviet sentiments. For instance, in April 1970, Finland issued its first stamp honoring a foreign individual—Vladimir Lenin—symbolizing the extent of Soviet influence. The Finnish media elite and political leaders often aligned their views with Soviet preferences to maintain favorable relations, suppressing anti-Soviet sentiments and banning literature that was deemed hostile to Soviet interests.

Despite these pressures, Finland retained its capitalist economic system, distinguishing itself from other Soviet-bordering nations adopting communist economies. This economic autonomy allowed Finland to engage in closer cooperation with other Nordic countries, fostering regional solidarity and further reinforcing its neutral stance in superpower politics. The ascent of Mikhail Gorbachev to Soviet leadership in 1985 marked a turning point, as Finnish media gradually began to criticize the Soviet Union more openly. Gorbachev himself cited Finland as a model for Eastern European nations transitioning away from strict Soviet control, signaling a slow erosion of Finlandization practices.

In West Germany, particularly among conservative politicians like Bavarian Prime Minister Franz Josef Strauss, Finlandization was a cautionary tale. It exemplified how a powerful neighbor could subtly dictate the internal affairs of a smaller, ostensibly independent state. Western intelligence agencies and political commentators closely monitored Finland, fearing similar pressures could undermine the sovereignty of other free states. In response, the United States funded research institutes and media outlets to counteract pro-Soviet narratives in Finland, contributing to the intense espionage activities that permeated Finnish society during this period.

Despite the constraints imposed by Finlandization, Finland managed to navigate its geopolitical challenges effectively. By fostering robust economic ties with the Nordic region and maintaining a strong commitment to capitalism, Finland preserved a degree of independence that allowed it to thrive economically while avoiding direct confrontation with the Soviet Union. This strategic neutrality ensured Finland's survival during the tense years of the Cold War and demonstrated the possibility of maintaining sovereignty and economic prosperity amidst overwhelming external pressures.

The year 1968 marked a significant escalation in Cold War tensions with the Soviet-led invasion of Czechoslovakia, one of the most substantial military operations on European soil since World War II. This event was precipitated by the Prague Spring, a brief period of political liberalization and reform within Czechoslovakia initiated by Alexander Dubček, the First Secretary of the Communist Party of Czechoslovakia. Dubček's "Action Program" aimed to create "socialism with a human face," advocating for increased freedom of the press, speech, and movement, economic reforms emphasizing consumer goods, the introduction of a multiparty government, limitations on the secret police, and the potential withdrawal from the Warsaw Pact.

The reforms of the Prague Spring were met with alarm in Moscow, where Soviet Premier Leonid Brezhnev and other hardline leaders perceived these changes as a threat to the integrity of the Eastern Bloc and the global communist movement. On August 20, 1968, the Soviet Union, along with troops from other Warsaw Pact nations—including East Germany, Poland, Hungary, and Bulgaria—invaded Czechoslovakia to halt the reforms and reassert Soviet control. The invasion was swift and overwhelming, resulting in widespread suppression of dissent and the reinstatement of hardline communist rule.

The Soviet intervention sparked a significant wave of emigration, with an estimated 70,000 Czechs and Slovaks fleeing the country immediately following the invasion. Over the subsequent years, the total number of refugees reached approximately 300,000 as citizens sought refuge from the oppressive regime. The invasion also provoked strong international condemnation and protests, not only from Western European countries but also from fellow socialist states such as Yugoslavia, Romania, and China, highlighting the fractures within the communist bloc.

The international community viewed the invasion as a blatant violation of Czechoslovakia's sovereignty and an act of imperialistic aggression. Western nations, already wary of Soviet intentions, saw the invasion as evidence of the Soviet Union's unwillingness to tolerate any deviation from orthodox Marxist-Leninist doctrine within its sphere of influence. This event further solidified the perception of the Soviet Union as an aggressive superpower intent on maintaining strict control over Eastern Europe, thereby intensifying Cold War hostilities.

In the wake of the 1968 invasion of Czechoslovakia, Soviet Premier Leonid Brezhnev articulated a new foreign policy principle known as the Brezhnev Doctrine. Delivered during his speech at the Fifth Congress of the Polish United Workers' Party in September 1968, the doctrine asserted the Soviet Union's right to intervene in any socialist country where the leadership sought to deviate from Marxist-Leninist principles. Brezhnev declared:

"When forces that are hostile to socialism try to turn the development of some socialist country towards capitalism, it becomes not only a problem of the country concerned but a common problem and concern of all socialist countries."

The Brezhnev Doctrine effectively justified military force to preserve socialist governments and prevent the spread of capitalist or reformist ideologies within the Eastern Bloc. It was a direct response to the Prague Spring, signaling to all member states of the Warsaw Pact that any attempt to pursue political liberalization or economic reforms without Soviet approval would be met with military intervention.

The doctrine was rooted in Marxism-Leninism's perceived failures to maintain economic and political stability in countries like Poland, Hungary, and East Germany. These nations were grappling with economic stagnation and declining living standards, especially in contrast to the rapidly growing economies of Western Europe. The Brezhnev Doctrine was intended to quell nationalist and reformist movements that threatened the cohesion and uniformity of the socialist bloc, thereby reinforcing Soviet dominance over its satellite states.

The implications of the Brezhnev Doctrine were profound and far-reaching. It effectively nullified any notions of sovereignty within the Eastern Bloc, as member states were compelled to adhere strictly to Soviet directives to avoid intervention. This policy stifled political dissent and hindered the natural evolution of these nations, contributing to long-term economic and social challenges. Additionally, the doctrine exacerbated tensions between the Soviet Union and the West, underscoring its willingness to use force to maintain its sphere of influence, thereby heightening Cold War anxieties.

The Brezhnev Doctrine remained a cornerstone of Soviet foreign policy until the late 1980s, when Mikhail Gorbachev's glasnost (openness) and perestroika (restructuring) began dismantling the rigid controls of the past. Gorbachev's approach facilitated greater political freedom and economic reform within the Eastern Bloc, effectively rendering the Brezhnev Doctrine obsolete and paving the way for the eventual dissolution of Soviet influence in Eastern Europe.

The events of Finlandization, the invasion of Czechoslovakia, and the establishment of the Brezhnev Doctrine significantly influenced the trajectory of the Cold War from the early 1960s through the late 1970s. These developments underscored the complexities of international relations during this period, where ideological rigidity and the quest for political control often clashed with emerging desires for political freedom and economic reform.

Finlandization highlighted the challenges smaller nations faced in maintaining sovereignty amidst the pressures exerted by superpowers. The Soviet Union's ability to influence Finnish politics without direct control demonstrated the effectiveness of soft power and political maneuvering in extending geopolitical influence. This nuanced form of control contrasted with the more overt military interventions seen in Czechoslovakia, showcasing the varied strategies employed by the Soviet Union to preserve its sphere of influence.

The invasion of Czechoslovakia and the subsequent Brezhnev Doctrine reinforced the Soviet Union's commitment to maintaining a unified and ideologically consistent Eastern Bloc. These actions not only solidified Soviet dominance over Eastern Europe but also heightened tensions with the West, contributing to the arms race and the continuous push for technological and military superiority. In the context of the Space Race, these geopolitical maneuvers underscored the urgency for both superpowers to demonstrate their prowess not only on Earth but also in space to showcase their ideological and technological superiority.

Moreover, these events influenced the broader strategy of détente that emerged in the 1970s. Recognizing the limitations imposed by rigid ideological control and the economic challenges faced by the Soviet Union led to a gradual shift towards seeking strategic accommodations and reducing the immediate threat of nuclear confrontation. This shift was reflected in various arms control agreements and increased diplomatic engagements, which sought to stabilize international relations and mitigate the risks of direct conflict.

The Cold War era was marked not only by the intense rivalry between the United States and the Soviet Union but also by a series of significant conflicts and political upheavals across the Third World. These events, often fueled by ideological battles between capitalism and communism, reshaped nations and influenced global power dynamics.

During the presidency of Lyndon B. Johnson, the United States adopted a more aggressive approach in Latin America, a strategy often referred to as the "Mann Doctrine." This hardline stance aimed to prevent the spread of communism in the Western Hemisphere. In 1964, the Brazilian military, with tacit US support, orchestrated a coup that overthrew the democratically elected government of João Goulart. This marked the beginning of a series of US-backed interventions to maintain pro-American regimes in the region.

In April 1965, the United States deployed approximately 22,000 troops to the Dominican Republic amidst the Dominican Civil War. The US government justified the intervention as a necessary measure to prevent the emergence of a Cuban-style revolution. The Organization of American States (OAS) facilitated the deployment through the predominantly Brazilian Inter-American Peace Force. The provisional presidency of Héctor García-Godoy was established until the conservative Joaquín Balaguer won the 1966 presidential election, defeating the deposed Juan Bosch, whose supporters faced severe repression and violence at the hands of Dominican police and armed forces.

In Indonesia, General Suharto emerged as a pivotal figure in the mid-1960s. As a staunch anti-communist, Suharto seized control from President Sukarno, initiating the "New Order" regime. From 1965 to 1966, with support from the United States and other Western nations, Suharto's military orchestrated the mass killing of over 500,000 individuals associated with the Indonesian Communist Party and other leftist groups. Additionally, hundreds of thousands were imprisoned under brutal conditions. A clandestine CIA report later characterized these massacres as among the worst mass murders of the 20th century, comparable to the Soviet purges, Nazi atrocities, and Maoist campaigns. These actions significantly shifted the balance of power in Southeast Asia, solidifying US influence in the region during the Cold War.

Simultaneously, the Vietnam War escalated dramatically under Johnson's administration. To counter the communist National Front for the Liberation of South Vietnam (NLF) and their North Vietnamese allies, the US committed 575,000 troops to Southeast Asia. This massive military buildup aimed to support Ngô Đình Diệm's South Vietnamese government against the insurgency. However, the prolonged conflict strained the US economy and ignited widespread domestic opposition, culminating in large-scale anti-war protests. By 1972, mounting pressures led to the gradual withdrawal of American forces. The withdrawal proved premature, as North Vietnam ultimately succeeded in conquering South Vietnam in 1975. The fall of Saigon not only marked a significant military defeat for the United States but also tarnished its global reputation, demonstrating the vulnerability of a superpower against a much poorer nation.

The Middle East remained a critical battleground for Cold War tensions. Egypt, heavily reliant on Soviet arms and economic assistance, became a focal point of contention. The Soviet Union felt compelled to support Egypt during the 1967 Six-Day War and the subsequent War of Attrition against Israel, a key US ally. Although Egypt began shifting its allegiance from the Soviet Union to the United States in 1972 under President Anwar Sadat, the Soviets continued to back Egypt and Syria during the Yom Kippur War in 1973, while the US steadfastly supported Israel.

The Soviet influence in the region extended beyond Egypt, as evidenced by their close ties with communist South Yemen and the nationalist regimes of Algeria and Iraq. In 1972, Iraq signed a 15-year Treaty of Friendship and Cooperation with the Soviet Union, disrupting the US-sponsored security framework of the Arab Cold War. In retaliation, the United States covertly funded Kurdish rebels during the Second Iraqi–Kurdish War. Despite these efforts, the Kurds were defeated by 1975, resulting in the forced relocation of hundreds of thousands of Kurdish civilians. Additionally, the Soviet Union provided indirect support to the Palestine Liberation Organization (PLO), led by Yasser Arafat, further entrenching its influence in the Israeli–Palestinian conflict.

In East Africa, the Ogaden War (1977-1978) highlighted the complexities of Cold War alliances. A territorial dispute between Somalia and Ethiopia over the Ogaden region led to Somali forces invading Ethiopia in June 1977. Both nations were Soviet clients at the time, with Somalia under the Marxist leadership of Siad Barre and Ethiopia governed by the pro-Soviet Derg regime led by Mengistu Haile Mariam. Initially, the Soviet Union attempted to mediate, but escalating tensions prompted Barre to sever ties with Moscow in November 1977, seeking support from China and the pro-American Safari Club—a coalition of intelligence agencies from Iran, Egypt, and Saudi Arabia.

Despite Somalia's shift, the Soviet Union remained committed to supporting Ethiopia, providing strategic planning and sophisticated arms. Cuban troops, numbering over a thousand by November 1975, played a crucial role in Ethiopia's successful counteroffensive against Somali forces. This intervention underscored the Soviet Union's strategic interests in maintaining influence in the Horn of Africa despite regional complexities.

Meanwhile, in Angola, the collapse of Portugal's Estado Novo regime during the Carnation Revolution of April 1974 led to the rapid decolonization of its African territories. Angola descended into a brutal civil war as three factions vied for power: the Marxist People's Movement for the Liberation of Angola (MPLA), the anti-communist National Union for the Total Independence of Angola (UNITA), and the National Liberation Front of Angola (FNLA). The MPLA, aligned with the Soviet Union, received substantial arms, while the CIA and China covertly supported the FNLA and UNITA. In response, Cuba intervened decisively in support of the MPLA, deploying over a thousand troops by November 1975. The combined efforts of Cuban forces and Soviet-supplied weaponry enabled the MPLA to secure victory, effectively quelling attempts by Zairean and South African troops to support their rivals.

The 1973 Chilean coup d'état further destabilized the political landscape of Latin America. Salvador Allende, a Marxist candidate from the Socialist Party, had been democratically elected as President of Chile in 1970, marking the first instance of a Marxist leader ascending to power in the Americas. However, the Central Intelligence Agency (CIA) orchestrated efforts to undermine Allende's government, fostering domestic unrest that culminated in a military coup led by General Augusto Pinochet in September 1973. Pinochet swiftly established a military dictatorship, reversing Allende's economic reforms and brutally suppressing leftist opposition through the Dirección de Inteligencia Nacional (DINA), which conducted widespread detentions, killings, and torture.

The international response to Pinochet's regime was mixed. Socialist states, excluding China and Romania, severed diplomatic ties with Chile. At the same time, the United States tacitly supported Operation Condor—a coordinated campaign of assassination and state terrorism executed by right-wing military dictatorships across the Southern Cone of South America. This operation aimed to eliminate leftist dissidents and was covertly backed by the US government, further entrenching authoritarianism in the region.

In Southeast Asia, Cambodia became another tragic theater of Cold War conflicts. During the Vietnam War, North Vietnamese forces established military bases in Cambodia, with the tacit approval of the Cambodian head of state Norodom Sihanouk, who sought to maintain neutrality. However, in March 1970, General Lon Nol deposed Sihanouk and aligned Cambodia with the United States, demanding the removal of North Vietnamese troops. This led to increased tensions and violence, as American and South Vietnamese forces launched bombing campaigns and ground incursions into Cambodia. The extensive US carpet bombing, which continued until 1973, aimed to disrupt the North Vietnamese supply lines but also devastated Cambodian rural society, exacerbating social divisions and causing tens of thousands of civilian casualties.

Amidst this chaos, the Khmer Rouge, a radical communist faction led by Pol Pot, gained momentum. Following Lon Nol's ousting of Sihanouk, the Khmer Rouge intensified their campaign to overthrow the Cambodian government. By 1975, they had seized Phnom Penh, instituting one of the most brutal regimes in history. Under Pol Pot's leadership, approximately 1.5 to 2 million Cambodians perished due to forced labor, starvation, and execution in what became known as the Killing Fields—a tragic testament to the extremes of ideological purges during the Cold War.

The international community's response to the Khmer Rouge was inconsistent. While the Soviet Union eventually supported the Kampuchean United Front for National Salvation, aiding Vietnam in its invasion of Cambodia in December 1978, the Khmer Rouge maintained a presence in the United Nations General Assembly, bolstered by backing from China, Western powers, and ASEAN member countries. This diplomatic impunity allowed the Khmer Rouge to persist in guerrilla warfare from refugee camps along the Thai border, prolonging Cambodia's recovery and destabilizing the region further.

The late 1960s and early 1970s were pivotal years in the Cold War, marked by significant shifts in alliances and diplomatic strategies. Central to this period was the Sino-Soviet split, a profound ideological and geopolitical rift between the Soviet Union and the People's Republic of China. This split not only redefined the dynamics within the communist bloc but also opened new avenues for the United States to recalibrate its foreign policy.

The Sino-Soviet split reached its zenith in 1969, culminating in heightened tensions along the extensive Chinese–Soviet border. This deterioration was rooted in deep-seated ideological differences and competing national interests. The split effectively fragmented the communist world, forcing both superpowers to navigate a complex landscape of shifting alliances and rivalries.

Amidst the turmoil of the Sino-Soviet split, United States President Richard Nixon recognized an unprecedented opportunity to alter the balance of power in the Cold War. Determined to exploit the discord between the Soviet Union and China, Nixon embarked on a policy of rapprochement with China, aiming to forge a strategic partnership that would counterbalance Soviet influence.

In 1972, Nixon visited China, the first by a U.S. president in over two decades. This landmark trip, which included meetings with Chinese Premier Zhou Enlai and Chairman Mao Zedong, signaled a dramatic shift in U.S. foreign policy. The visit not only paved the way for improved Sino-American relations but also set the stage for the eventual normalization of diplomatic ties. This process culminated in 1979 with the signing of the Joint Communiqué on the Establishment of Diplomatic Relations between President Jimmy Carter and Chinese Premier Deng Xiaoping.

Parallel to the rapprochement with China, Nixon also sought to ease tensions with the Soviet Union through a policy known as détente. Following the ousting of Nikita Khrushchev, Leonid Brezhnev emerged as the new

General Secretary of the Communist Party of the Soviet Union. Under Brezhnev's leadership, the Soviet Union pursued a policy of collective leadership alongside Premier Alexei Kosygin and Chairman Nikolai Podgorny.

Nixon's administration engaged in a series of high-level negotiations with Brezhnev, leading to significant arms control agreements. The Strategic Arms Limitation Talks (SALT) produced two landmark treaties: SALT I and the Anti-Ballistic Missile (ABM) Treaty. SALT I, signed in 1972, was the first comprehensive limitation pact between the superpowers, aiming to curb the arms race by restricting the number of strategic ballistic missiles. The ABM Treaty, signed in 1972 also, prohibited the development of systems designed to intercept incoming ballistic missiles, thereby limiting the escalation of nuclear arsenals.

These treaties began a new era of "peaceful coexistence," where the United States and the Soviet Union sought to manage their rivalry through dialogue and negotiation rather than confrontation. Détente also encompassed efforts to strengthen economic ties, with agreements signed between 1972 and 1974 to enhance trade and reduce economic barriers.

In 1975, the Conference on Security and Co-operation in Europe culminated in the signing of the Helsinki Accords. This significant agreement sought to stabilize the situation in Europe by addressing a range of political, economic, and security issues. One of the most contentious aspects of the Accords was the inclusion of human rights provisions. The Soviet Union pledged to respect human rights and fundamental freedoms, including the right to self-determination and the protection of property rights.

However, in practice, the Soviet government often disregarded these commitments. Human rights activists within the USSR faced regular harassment, repression, and arrest, as the authorities prioritized maintaining control over adhering to the principles outlined in the Accords. Despite this, the Helsinki Accords provided a framework for future human rights advocacy and became a reference point for dissidents seeking to challenge the Soviet regime.

Amid these geopolitical shifts, American business magnate Armand Hammer played a crucial role in facilitating trade between the United States and the Soviet Union. As the head of Occidental Petroleum, Hammer leveraged his extensive business network to act as an intermediary between the two superpowers. His efforts were instrumental in establishing and maintaining economic ties during intense political rivalry.

Hammer's relationship with the Soviet leadership spanned decades, beginning in the 1920s with Vladimir Lenin's endorsement. Despite interruptions during Stalin's rise to power, Hammer advocated for East-West trade. By 1974, Soviet General Secretary Leonid Brezhnev publicly acknowledged Hammer's contributions, recognizing his role in bridging the economic divide between the two nations. By 1981, Hammer had become personally acquainted with Brezhnev, illustrating the significant impact of his diplomatic and business endeavors.

President Nixon and his National Security Advisor, Henry Kissinger, embodied a realist approach to foreign policy, prioritizing pragmatic strategies over ideological commitments. Rejecting the traditional Cold War emphasis on anti-communism and the promotion of democracy, Nixon and Kissinger sought to achieve peace, economic stability, and cultural exchanges. They recognized that the United States could no longer sustain extensive military commitments without jeopardizing its economic health and societal well-being.

This shift towards realpolitik involved downsizing America's global engagements and focusing on achievable diplomatic objectives. Nixon and Kissinger's policies reflected a recognition that the idealistic goals of containing communism were increasingly untenable in the face of economic constraints and shifting public opinion. Their approach emphasized practical solutions and strategic alliances, laying the groundwork for subsequent foreign policy directions.

The détente policies and strategic arms limitation treaties represented significant milestones in Cold War diplomacy. While détente sought to reduce direct confrontation and promote stability, it also revealed the inherent tensions and limitations within the U.S.-Soviet relationship. The Helsinki Accords, despite their flawed implementation, provided a platform for future human rights advocacy and international cooperation.

Ultimately, the détente era set the stage for the eventual thaw in Cold War hostilities, contributing to the conditions that would later facilitate the end of the Cold War. The strategic realignments and diplomatic initiatives of the 1970s demonstrated the potential for peaceful coexistence and mutual understanding, even amidst profound ideological differences.

The late 1970s marked a critical turning point in Cold War dynamics, as the initial thaw of détente gave way to renewed hostility and competition. The Soviet invasion of Afghanistan shattered the fragile peace established during the earlier decades, reigniting Cold War tensions and setting the stage for a more aggressive and confrontational phase of global politics. This period of intensified rivalry reshaped international relations and had profound implications for the political landscapes of numerous Third World countries caught in the crossfire of superpower competition. The legacy of these developments continued to influence the trajectory of the Cold War, ultimately contributing to the eventual dissolution of the Soviet Union and the redefinition of global power structures in the subsequent decades.

As the 1970s progressed, the fragile détente between the United States and the Soviet Union began to unravel, signaling a return to heightened Cold War tensions. This period was marked by a resurgence of ideological confrontations, geopolitical conflicts, and internal struggles within both superpowers, culminating in the Soviet invasion of Afghanistan and the onset of the "New Cold War."

During the late 1970s, the Soviet Union intensified its efforts to suppress internal dissent. Under the leadership of Yuri Andropov, the KGB ramped up its persecution of prominent Soviet dissidents, including renowned writers and intellectuals like Aleksandr Solzhenitsyn and Andrei Sakharov. These individuals openly criticized the Soviet regime, challenging its policies and advocating for greater political freedoms and human rights. Their harsh treatment underscored the Soviet government's commitment to maintaining strict control over ideological conformity, even as external relations with the West gradually improved.

Despite the era of détente, indirect conflicts between the superpowers persisted, particularly in the developing nations of the Third World. Political crises and civil wars in regions such as the Middle East, Chile, Ethiopia, and Angola became battlegrounds for proxy wars, where the United States and the Soviet Union supported opposing factions to extend their influence. These interventions often exacerbated local conflicts, leading to prolonged instability and suffering for the populations involved. The continuation of these indirect confrontations highlighted the enduring competition for global dominance, even amidst attempts to ease direct tensions between the superpowers.

In 1973, President Richard Nixon announced his administration's commitment to pursuing most-favored-nation (MFN) trade status with the Soviet Union, a policy aimed at fostering economic cooperation despite ongoing political differences. However, this initiative faced significant challenges in the United States Congress, leading to the introduction of the Jackson-Vanik Amendment in 1974. The amendment linked the granting of MFN trade status to the Soviet Union with the United States' foreign policy objectives, particularly concerning the rights of persecuted Soviet Jews to emigrate.

The Jackson-Vanik Amendment effectively restricted the President's ability to extend MFN status to the USSR unless the Soviet government allowed the free emigration of Jewish refuseniks—individuals denied permission to leave the Soviet Union due to their Jewish heritage. This legislative move underscored the growing importance of human rights issues in U.S. foreign policy and highlighted the limitations of economic engagement when fundamental human rights were at stake.

Amidst these tensions, efforts to control the arms race continued. In 1979, President Jimmy Carter sought to advance arms limitation through the Strategic Arms Limitation Talks II (SALT II). This treaty aimed to curtail the proliferation of nuclear weapons by setting limits on the number of strategic ballistic missiles and other nuclear delivery systems each superpower could maintain. However, the progress towards SALT II was impeded by escalating geopolitical crises, including the Iranian Revolution and the Nicaraguan Revolution, which saw the overthrow of pro-U.S. governments.

Moreover, the Soviet Union's increasing involvement in Afghanistan further strained relations. The SALT II negotiations ultimately failed to achieve their objectives, as the Soviet intervention in Afghanistan overshadowed diplomatic efforts and heightened mutual distrust between the United States and the Soviet Union.

The culmination of these deteriorating relations occurred in December 1979, when the Soviet Union launched a full-scale invasion of Afghanistan. The invasion, known as Operation Storm-333, aimed to overthrow the newly established government of Hafizullah Amin, a pro-KGB leader who had seized power through internal coup. Soviet special forces swiftly captured the Tajbeg Palace in Kabul, eliminating Amin and installing Babrak Karmal, a more compliant leader aligned with Soviet interests.

This military intervention marked a decisive end to the period of détente, reigniting Cold War hostilities. The United States responded vehemently to the Soviet invasion, viewing it as a direct threat to regional stability and a violation of Afghan sovereignty. President Carter condemned the invasion as "the most serious threat to peace since World War II" and implemented a series of punitive measures against the Soviet Union. These measures included withdrawing the SALT II treaty from ratification, imposing embargoes on grain and technology shipments, and demanding a significant increase in U.S. military spending.

In addition to these economic sanctions, the United States, alongside 65 other nations, boycotted the 1980 Summer Olympics in Moscow as a symbolic protest against the Soviet aggression. This multifaceted response underscored the deepening rift between the superpowers and the United States' commitment to countering Soviet expansionism.

The Soviet invasion of Afghanistan is widely regarded as the catalyst for the "New Cold War," a period characterized by renewed and intensified rivalry between the United States and the Soviet Union. This phase saw both superpowers adopting more militant stances and escalating their efforts to undermine each other's influence globally.

Under President Ronald Reagan, the New Cold War reached new heights of antagonism. Reagan's administration pursued a more confrontational approach, exemplified by the support of counterinsurgencies and anti-communist movements in various Third World countries. The Reagan Doctrine emphasized the rollback of Soviet influence and the promotion of democracy and free-market principles, contrasting sharply with the détente policies of previous administrations.

Simultaneously, the Soviet Union faced significant internal challenges, including economic stagnation and a burgeoning military budget that strained its resources. The Soviet economy, heavily burdened by extensive military expenditures—comprising 40–60% of the federal budget and up to 15% of GDP—struggled to sustain both its domestic needs and its international commitments. These economic pressures, coupled with the ongoing war in Afghanistan, further weakened the Soviet Union's global standing and exacerbated internal dissent.

President Ronald Reagan and Prime Minister Margaret Thatcher emerged as central figures in the late Cold War period, steering the United States and the United Kingdom towards a more confrontational stance against the Soviet Union. Their unwavering commitment to defeating communism through increased military spending, strategic economic measures, and support for anti-communist movements worldwide marked a significant shift in Cold War dynamics. The legacy of Reagan and Thatcher's policies played a crucial role in shaping the geopolitical landscape of the late 20th century, ultimately contributing to the resolution of Cold War tensions and the redefinition of global power structures.

The late 1970s and early 1980s witnessed a significant shift in the dynamics of the Cold War, characterized by the emergence of staunchly anti-communist leaders in the United States and the United Kingdom. President Ronald Reagan of the United States and Prime Minister Margaret Thatcher of the United Kingdom became pivotal figures in redefining Western strategies against the Soviet Union, ushering in an era marked by heightened tensions and assertive policies to curb Soviet influence globally.

Ronald Reagan, a former actor and Governor of California, entered the 1980 presidential race with a clear and uncompromising vision for America's role in the Cold War. In a candid conversation with Richard V. Allen in January 1977, Reagan articulated his straightforward approach to dealing with the Soviet Union: "We win and they lose." This declaration underscored his belief in the inevitability of American triumph over Soviet communism and set the tone for his future policies.

Elected in November 1980, Reagan swiftly moved to implement his vision. He vowed to significantly increase military spending, aiming to outpace Soviet capabilities and force the USSR into an untenable position. This aggressive stance was encapsulated in his characterization of the Soviet Union as an "evil empire," a term that resonated deeply with the American public and signaled a departure from the more conciliatory approaches of previous administrations.

Reagan's policies were not merely rhetorical. In 1982, he attempted to undermine the Soviet economy by obstructing Moscow's proposed gas pipeline to Western Europe. This move was intended to cut off a crucial source of hard currency for the USSR, thereby exacerbating its economic difficulties. Although the initiative ultimately faltered due to resistance from American allies in Europe who depended on Soviet gas revenues, it exemplified Reagan's willingness to take bold actions to challenge Soviet economic strategies.

Margaret Thatcher, who became Prime Minister of the United Kingdom in 1979, quickly established herself as a formidable ally of Reagan. Known as the "Iron Lady" for her unwavering resolve, Thatcher shared Reagan's vehement opposition to Soviet communism. She echoed his sentiments by labeling the Soviet Union as "bent on world dominance," reinforcing the narrative of a clear ideological adversary.

Thatcher's policies complemented Reagan's approach, particularly in the realms of economic and military strategy. She advocated for free-market principles and reduced government intervention, aligning with Reagan's economic policies to strengthen Western economies and counterbalance Soviet power. Together, Reagan and Thatcher fostered a robust Anglo-American alliance that became a cornerstone of Western resistance against Soviet expansionism.

By the mid-1980s, Reagan's anti-communist stance had evolved into what became known as the Reagan Doctrine. This strategic framework extended beyond mere containment of Soviet influence, advocating for the active subversion and overthrow of existing communist governments. The Reagan Doctrine marked a significant escalation in U.S. foreign policy, emphasizing support for anti-communist insurgencies and movements worldwide.

Under this doctrine, the United States substantially aided various resistance groups fighting Soviet-backed regimes. Notably, the CIA supported the mujahideen in Afghanistan, who were engaged in a fierce struggle against the Soviet invasion. Additionally, the Reagan administration encouraged the Pakistan Inter-Services Intelligence (ISI) to train and equip fighters from Muslim-majority regions to resist Soviet control, thereby extending the conflict into new theaters.

Thatcher supported these initiatives, recognizing the strategic importance of weakening Soviet influence through indirect means. Her government facilitated intelligence cooperation and provided diplomatic backing for Reagan's policies, reinforcing the transatlantic commitment to opposing Soviet expansion.

The influence of Reagan and Thatcher was also evident in their responses to political developments in Eastern Europe. The rise of the Solidarity movement in Poland, led by Lech Wałęsa and inspired by the moral authority of Pope John Paul II, presented a significant challenge to Soviet dominance in the region. Solidarity, a powerful trade union, galvanized widespread opposition against the communist regime, embodying the spirit of resistance that Reagan and Thatcher championed.

In December 1981, faced with increasing unrest and the potential collapse of the communist government, Polish leader Wojciech Jaruzelski declared martial law. Reagan responded by imposing economic sanctions on Poland, further isolating the Soviet-backed regime and demonstrating unwavering support for democratic movements. These actions underscored the commitment of both leaders to fostering political change in Eastern Europe, aligning with their broader strategy to undermine Soviet influence.

Reagan and Thatcher's collaborative efforts significantly intensified Cold War pressures on the Soviet Union. Their policies not only strained Soviet resources but also emboldened dissident movements within the Eastern Bloc, contributing to the eventual decline of Soviet power. The combination of increased military expenditure, strategic economic initiatives, and unwavering support for anti-communist movements created a multifaceted challenge that the Soviet Union struggled to manage.

Their leadership also fostered a renewed sense of Western unity and purpose, galvanizing support for democratic values and free-market principles. This period of intensified rivalry, often referred to as the "Second Cold War," laid the groundwork for the eventual dissolution of the Soviet Union in 1991 and the end of the Cold War era.

The twilight years of the Cold War were marked by escalating military competition and profound economic strains that ultimately contributed to the unraveling of the Soviet Union. This period witnessed a relentless arms race, strategic defense initiatives, significant foreign interventions, and critical incidents that heightened tensions between the two superpowers, shaping the geopolitical landscape of the late 20th century.

Throughout the 1970s, the Soviet Union devoted an immense portion of its gross national product (GNP) to its military apparatus, with defense expenditures accounting for up to 25 percent of the economy. This substantial allocation was driven not by genuine military necessity but largely by the interests of the nomenklatura—the Soviet elite whose power and privileges were tied to the military-industrial complex. Consequently, the Soviet Armed Forces burgeoned into the largest in the world, boasting vast arsenals of nuclear and conventional weapons, a colossal number of troops, and an extensive military-industrial base.

However, this military expansion came at a severe cost. The relentless pursuit of military superiority exacerbated existing structural deficiencies within the Soviet economy, leading to a prolonged period of economic stagnation known as the Era of Stagnation during the late Brezhnev years. Investment in consumer goods and civilian sectors was severely neglected, resulting in a decline in living standards and widespread public dissatisfaction. Despite the quantitative advantages of the Soviet military, qualitative deficiencies were starkly evident. For instance, during the Persian Gulf War, the Soviet T-72 main battle tank was found to be technologically inferior to the American M1 Abrams, despite the USSR deploying nearly three times as many T-72s as the US fielded M1s.

In response to the perceived Soviet threat, President Ronald Reagan initiated the most substantial peacetime military buildup in American history. Between 1981 and 1986, military spending increased from 5.3 percent to 6.5 percent of GNP, reflecting Reagan's commitment to outpacing Soviet military capabilities. This aggressive stance was encapsulated in Reagan's characterization of the Soviet Union as an "evil empire," a term that resonated deeply with the American public and signaled a departure from the more conciliatory approaches of previous administrations.

One of Reagan's most ambitious projects was the Strategic Defense Initiative (SDI), popularly dubbed "Star Wars" by the media. Announced in 1983, SDI aimed to develop a missile defense system capable of intercepting and destroying incoming ballistic missiles in mid-flight, thereby neutralizing the Soviet nuclear threat. While technologically visionary, SDI faced substantial criticism for its feasibility and potential to escalate the arms race further.

The escalation of the arms race saw both superpowers deploying increasingly sophisticated weaponry. The Soviet Union responded to the US missile deployments by introducing the RSD-10 Pioneer ballistic missiles targeting

Western Europe. NATO, under the impetus of the Carter presidency, deployed the MGM-31 Pershing missiles and cruise missiles primarily in West Germany. These deployments placed Soviet territories within a mere ten minutes' striking distance of potential American strikes, heightening the immediacy of the threat and deepening Cold War tensions.

Despite the immense military buildup, the Soviet economy could not sustain the heavy financial burdens. Inefficient command economics, reduced oil revenues due to increased production by Saudi Arabia and other non-OPEC nations, and declining oil prices contributed to a severe economic downturn. Oil was the backbone of the Soviet export economy, and the 1980s oil glut severely impacted Soviet revenues. Coupled with inefficient planned manufacturing and collectivized agriculture, these factors led to widespread economic stagnation, undermining the very foundations of the Soviet system.

Several critical incidents in the early 1980s significantly heightened US-USSR tensions. On September 1, 1983, the Soviet Union shot down Korean Air Lines Flight 007, a Boeing 747 passenger jet carrying 269 people, including Congressman Larry McDonald. The Soviet Air Force mistook the aircraft for a spy plane and destroyed it with air-to-air missiles. The incident, which resulted in the loss of 269 lives, was widely condemned and reinforced Reagan's hardline stance against the Soviet Union.

Another pivotal moment was the 1983 Soviet nuclear false alarm incident. On September 26, 1983, a system glitch at Serpukhov-15 falsely indicated that multiple intercontinental ballistic missiles were heading toward the USSR. Lieutenant Stanislav Petrov, a Soviet officer, correctly identified the alert as a false alarm and chose not to escalate the response, potentially averting a catastrophic nuclear war. Petrov's decision underscored the perilous nature of Cold War tensions and the risks of miscommunication between the superpowers.

The Able Archer 83 exercise in November 1983 further strained relations. This NATO military exercise, which simulated a coordinated nuclear release, was perceived by the Soviet leadership as a possible prelude to an actual nuclear strike. The heightened state of alert during this period was one of the most dangerous moments of the Cold War, reminiscent of the Cuban Missile Crisis. It underscored the fragility of the precarious peace.

During the late 1970s and early 1980s, the Reagan administration pursued an aggressive foreign policy to counter Soviet influence worldwide. This included support for counterinsurgency efforts in the Third World, exemplified by interventions in Lebanon, Grenada, Libya, and Central America. The support for the Contras in Nicaragua, aimed at overthrowing the Soviet-aligned Sandinista government, became particularly controversial, leading to the Iran-Contra affair.

Reagan's administration emphasized quick, low-cost interventions to avoid the prolonged quagmire experienced in Vietnam and Afghanistan. However, these interventions often resulted in mixed outcomes and domestic controversies, reflecting the complex nature of Cold War geopolitics. The invasion of Grenada in 1983 and the bombing of Libya in 1986 were popular among many Americans but sparked significant international debate and criticism.

The Soviet Union's involvement in Afghanistan, launched in December 1979, was a critical factor in the deterioration of US-USSR relations. Operation Storm-333 saw Soviet special forces overthrowing President Hafizullah Amin and installing Babrak Karmal as the new leader. This intervention sparked a fierce guerrilla war led by the mujahideen, who received substantial support from the United States, Pakistan, China, and other allies. The protracted conflict drained Soviet resources and contributed to the image of the USSR as entangled in a costly and unwinnable war, akin to the American experience in Vietnam.

The Afghan war not only strained the Soviet economy but also intensified domestic dissatisfaction with the government, accelerating the decline of the Soviet Union. The conflict demonstrated the limits of Soviet military power and further eroded the legitimacy of the Soviet regime. The protracted nature of the war, coupled with growing economic woes, highlighted the unsustainable nature of the Soviet military and economic policies.

Economic policies and external factors played a significant role in the late Cold War dynamics. The Soviet Union's command economy struggled to keep pace with the technological advancements and economic growth of the West. Issues such as inefficient resource allocation, corruption, and lack of innovation hindered economic progress, while declining oil prices in the early 1980s reduced vital export revenues.

Externally, the global oil glut of the 1980s, driven by increased production from non-OPEC nations and Saudi Arabia, severely impacted Soviet export earnings. As oil was the main source of Soviet foreign currency, the glut exacerbated economic difficulties, limiting the USSR's ability to sustain its military and foreign commitments. Additionally, Saudi Arabia's strategic increase in oil production further destabilized global oil markets, placing additional pressure on the Soviet economy.

Amid these geopolitical tensions, individual stories highlighted the Cold War's human dimension. Ten-year-old American Samantha Smith wrote a letter to Soviet leader Yuri Andropov, expressing her fear of nuclear war and advocating for peace. In response, Andropov invited Smith to the Soviet Union, an unprecedented gesture symbolizing a fleeting moment of hope for better relations. Smith's visit humanized the adversarial relationship, demonstrating that amidst the political and military confrontations, there remained a desire for understanding and peace among ordinary people.

The late 1970s and early 1980s marked a period of escalating tensions and intensified rivalry between the United States and the Soviet Union. Reagan's military buildup, the Strategic Defense Initiative, and aggressive foreign interventions, combined with the Soviet Union's economic struggles and costly military engagements, created unsustainable pressures on both superpowers. Key incidents such as the downing of Korean Air Lines Flight 007 and the nuclear false alarm underscored the perilous nature of the Cold War during this era.

Ultimately, the combination of economic inefficiencies, internal dissent, and relentless external pressures contributed to the weakening of the Soviet Union, setting the stage for its eventual collapse in 1991. This period's intense military and economic competition played a pivotal role in shaping the final chapters of the Cold War, highlighting the profound impact of military strategy and economic policy on global geopolitical dynamics. The legacy of these developments not only reshaped international relations but also paved the way for a new world order in the post-Cold War era.

The final years of the Cold War were characterized by transformative political reforms within the Soviet Union, groundbreaking diplomatic engagements between superpowers, and a series of pivotal events that ultimately led to the dissolution of the Soviet Union and the end of decades-long global tension. This period, spanning from 1985 to 1991, witnessed a remarkable shift from entrenched rivalry to unprecedented cooperation, driven largely by the visionary leadership of Soviet General Secretary Mikhail Gorbachev and the strategic diplomacy of U.S. President Ronald Reagan.

In March 1985, a relatively young Mikhail Gorbachev assumed the role of General Secretary of the Communist Party of the Soviet Union, inheriting an economy teetering on the brink of collapse. The Soviet Union was grappling with severe economic stagnation, exacerbated by a sharp decline in oil prices during the 1980s, significantly reducing foreign currency earnings—the lifeblood of the Soviet economy. Faced with these daunting challenges, Gorbachev recognized that superficial measures would be insufficient to revive the ailing state. Instead, he embarked on a path of profound structural transformation.

In June 1987, Gorbachev introduced Perestroika, meaning "restructuring," an ambitious agenda to revitalize the Soviet economy. Perestroika sought to dismantle the rigid production quota system that had long hindered economic efficiency and innovation. Perestroika intended to redirect resources from the costly arms race toward more productive civilian sectors by allowing cooperative ownership of small businesses and opening the door to foreign investment. These reforms were designed to stimulate economic growth and reduce the burdens imposed by excessive military expenditures.

Simultaneously, Gorbachev launched Glasnost, or "openness," a policy to increase transparency within state institutions and promote freedom of the press. Glasnost was intended to mitigate corruption within the Communist Party and curb the abuse of power by party officials. By fostering an environment of greater political and social openness, Glasnost enabled Soviet citizens to engage more freely with the Western world, thereby facilitating an accelerating détente between the Soviet Union and the United States.

Gorbachev's reforms coincided with a significant thaw in relations between the United States and the Soviet Union. Responding to the Kremlin's willingness to engage in dialogue and reduce military tensions, President Ronald Reagan initiated a series of high-level summits aimed at scaling back the arms race and fostering economic cooperation. The first of these summits occurred in November 1985 in Geneva, Switzerland, setting the stage for subsequent diplomatic breakthroughs.

A pivotal moment in this rapprochement occurred during the Reykjavík Summit in October 1986 in Iceland. Although initial discussions faltered over Reagan's proposed Strategic Defense Initiative (SDI), colloquially known as "Star Wars," the summit laid the groundwork for future negotiations. The third summit, held in Washington, D.C., in December 1987, proved to be a turning point. Here, Gorbachev and Reagan signed the Intermediate-Range Nuclear Forces (INF) Treaty. This landmark agreement eliminated all nuclear-armed, ground-launched ballistic and cruise missiles with ranges between 500 and 5,500 kilometers. The INF Treaty not only marked the first instance of the elimination of an entire class of nuclear weapons but also symbolized the tangible progress toward reducing the specter of nuclear war.

In June 1987, President Reagan delivered one of the most iconic speeches of the Cold War era at the Brandenburg Gate in Berlin. Standing before the imposing Berlin Wall, Reagan famously urged Soviet leader Mikhail Gorbachev to "tear down this wall!" This impassioned plea encapsulated the growing demand for freedom and the reunification of East and West Germany. Reagan's speech resonated deeply with the people of Berlin and the broader international community, further pressuring the Soviet Union to reconsider its rigid stance in Eastern Europe.

The late 1980s saw a series of seismic shifts in Eastern Europe. In 1989, the Soviet Union made a decisive move by declaring it would no longer intervene in the affairs of its satellite states in Central and Eastern Europe. This policy shift empowered popular movements seeking greater autonomy and democratic reforms. The most symbolic outcome of this change was the fall of the Berlin Wall on November 9, 1989. This momentous event signaled the imminent collapse of communist regimes across the region and the reunification of Germany.

The Soviet Union's protracted involvement in Afghanistan had long been a drain on its resources and morale, often likened to the American experience in Vietnam. By 1989, the untenable costs and the unyielding resistance from the mujahideen, supported covertly by the United States and other allies, compelled the Soviet leadership to withdraw its forces. The final Soviet troops left Afghanistan in February 1989, marking the end of a costly and unpopular conflict that had significantly weakened the Soviet Union's international standing and internal cohesion.

In December 1989, Gorbachev and U.S. President George H. W. Bush formally declared the Cold War over in the Malta Summit. This declaration was followed by the signing of the Treaty on the Final Settlement with Respect to Germany in September 1990, which paved the way for German reunification. The collapse of the Berlin Wall had already set the stage for the dissolution of East Germany, and the treaty underscored the complete normalization of relations between Germany and its neighbors.

Collaborative Efforts and Final Agreements

The relationship between Gorbachev and President Bush continued to evolve, culminating in the START I (Strategic Arms Reduction Treaty) signed in July 1991. This treaty aimed to significantly reduce the number of strategic nuclear weapons held by both superpowers, further diminishing the nuclear threat and solidifying the post-Cold War security architecture.

Additionally, the two nations cooperated in addressing regional conflicts, exemplified by their joint participation in the Gulf War (August 1990 – February 1991). This collaboration demonstrated a new era of partnership, where former adversaries worked together to confront mutual threats, further eroding the foundations of Cold War antagonism.

Despite these diplomatic successes, the Soviet Union was grappling with insurmountable internal challenges. The combination of economic inefficiencies, political unrest, and a loss of ideological cohesion rendered the Soviet state increasingly untenable. On December 25, 1991, Mikhail Gorbachev resigned as President of the Soviet Union, and the following day, the Soviet Union was officially dissolved. The Commonwealth of Independent States (CIS) emerged from the remnants of the Soviet Union, marking the definitive end of the Cold War.

The decade following the height of the Space Race was marked by profound geopolitical shifts that ultimately led to the dissolution of the Soviet Union and the transformation of Eastern Europe. Central to this era were the weakening economic and political structures of the Soviet Union and the various, often fragmented, attempts to implement reforms aimed at reversing this decline.

Kenneth S. Deffeyes, in his analysis, posited that the Reagan administration strategically encouraged Saudi Arabia to reduce oil prices. This maneuver aimed to undermine the Soviet economy by making it unprofitable for them to sell oil, thereby depleting the USSR's hard currency reserves. The leadership instability within the Soviet Union further exacerbated the economic strain. After the long tenure of Leonid Brezhnev, his successors Yuri Andropov and Konstantin Chernenko, both deeply rooted in Brezhnev's traditionalist approach, held power briefly due to their advanced ages—Andropov died at 68 and Chernenko at 72, each serving less than two years.

In a decisive move to break the cycle of short-lived leadership, the Soviet Union appointed Mikhail Gorbachev in 1985. Gorbachev introduced significant economic reforms under the banner of perestroika and promoted greater transparency through glasnost, dismantling decades of stringent government censorship. These policies not only revitalized the Soviet economy but also fostered an environment of openness that challenged the existing political order. Gorbachev's vision extended beyond internal reforms; he sought to de-escalate the Cold War tensions, leading to the Soviet withdrawal from Afghanistan in 1988 and a strategic reduction of military involvement in Eastern Europe.

The pivotal moment came in 1989 when Gorbachev signaled non-interference in the internal affairs of Soviet satellite states. This policy shift emboldened grassroots movements across Eastern Europe, most notably Poland's Solidarity movement, which rapidly gained widespread support. The Pan-European Picnic held in August 1989 on the Hungarian-Austrian border catalyzed change. Organized by Otto von Habsburg and Hungarian Minister of State Imre Pozsgay, the event tested Gorbachev's response to mass emigration from East Germany. Thousands of East Germans, inspired by the open border, flocked to Hungary, effectively breaching the Iron Curtain without military intervention. This peaceful exodus revealed the fragility of the communist regimes and demonstrated that the Soviet Union was no longer willing or able to enforce its control with force.

The consequences of these developments were swift and far-reaching. By 1989, the Soviet alliance system was on the verge of collapse. The Warsaw Pact states, stripped of Soviet military backing, saw their communist leaders increasingly lose authority. In Poland and Hungary, governments were the first to negotiate the organization of competitive elections, setting a precedent for democratic transitions. In Czechoslovakia and East Germany, mass protests successfully ousted entrenched communist leaders. Bulgaria and Romania followed suit, with Romania experiencing a particularly violent uprising that led to the execution of its head of state, a stark departure from the otherwise predominantly peaceful transitions in the region.

The symbolic culmination of this wave of change occurred with the fall of the Berlin Wall in November 1989. This event not only signified the physical dismantling of the Iron Curtain but also represented the ideological collapse of European communist governments. The Berlin Wall's demise facilitated the reunification of East and

West Germany and underscored the irreversible shift towards democracy and market economies in Central and Eastern Europe.

Simultaneously, the internal dynamics of the Soviet Union were unraveling. The Soviet republics began asserting their sovereignty, invoking Article 72 of the USSR constitution to declare independence. In April 1990, a law was enacted allowing republics to secede following a two-thirds majority vote in a referendum. This legal framework empowered republics like Lithuania, which declared its independence after a resounding victory for the Sąjūdis movement in March 1990, challenging the legitimacy of Soviet occupation. Attempts by Soviet forces to suppress these movements through violent means, such as the events known as Bloody Sunday in Lithuania and the Barricades in Latvia, only intensified international support for the secessionists.

The final blow to the Soviet Union came in August 1991 during the attempted coup by hardline government officials and members of the KGB who sought to reverse Gorbachev's reforms and restore central control. The coup's failure significantly weakened Gorbachev's authority and elevated Boris Yeltsin, the President of Russia, to prominence. Yeltsin's decisive actions during the coup endeared him to the public and shifted the balance of power towards the republics. Following the coup, Latvia and Estonia swiftly declared their full independence, followed by other republics.

By December 1991, the Soviet Union formally dissolved, giving rise to fifteen independent states. The Russian Federation emerged as the largest and most influential successor state, inheriting the Soviet Union's membership in the United Nations, its permanent seat on the Security Council, and control over its nuclear arsenal and military forces. The dissolution marked the end of an era and the beginning of a new geopolitical landscape, fundamentally altering the dynamics that had previously driven the Space Race.

In the aftermath, global leaders recognized the significance of this transformation. In his 1992 State of the Union Address, U.S. President George H. W. Bush acknowledged the end of the Cold War, stating, "By the grace of God, America won the Cold War." This sentiment was echoed in the burgeoning partnership between the United States and Russia, exemplified by the Strategic Arms Reduction Treaty (START II) agreed upon in 1993, which aimed to reduce nuclear arsenals further and solidify the new era of cooperation.

The collapse of the Soviet Union and the peaceful revolutions in Eastern Europe not only reshaped the world's political map but also had profound implications for the continuation and legacy of the Space Race. The end of the Cold War removed the intense rivalry that had driven much of the early space exploration efforts, paving the way for collaborative international endeavors in space, such as the International Space Station, and marking a new chapter in humanity's quest to explore beyond our planet.

The dissolution of the Soviet Union and the subsequent transformation of Eastern Europe ushered in a new era of global dynamics, reshaping international relations and national identities in profound and lasting ways. Vladislav Zubok aptly characterized this period as an event of "epochal geopolitical, military, ideological, and economic significance," highlighting the far-reaching consequences that extended well beyond the immediate political upheavals.

In the immediate aftermath of the Soviet collapse, Russia faced severe economic challenges. The abrupt reduction in military spending, a cornerstone of the Soviet economy, precipitated a drastic restructuring that left millions unemployed. Western analysts have compared the resulting recession in Russia during the early 1990s to the Great Depression experienced by the United States and Germany, noting that the economic turmoil was both deeper and more prolonged. Neoliberal reforms, intended to transition Russia towards a market economy, instead led to widespread hardship, with only a handful of post-communist states managing to integrate successfully into the global capitalist system. Most struggled to recover, with many remaining significantly behind their Western counterparts, a disparity that analysts predict may take several decades to bridge.

The Baltic states—Estonia, Latvia, and Lithuania—embarked on a unique path in asserting their independence. These nations viewed themselves as the revival of the pre-1940 independent states, arguing that their incorporation into the Soviet Union was both illegal and illegitimate. In 1990 and 1991, they reasserted their sovereignty, which was met with resistance from Soviet authorities who attempted to suppress these declarations through force. Events such as Bloody Sunday in Lithuania and the Barricades in Latvia, where Soviet forces clashed with civilians, only served to galvanize international support for the secessionist movements, ultimately solidifying their independence.

The process of decommunization varied significantly across the former Soviet sphere. In many Eastern European countries, communist parties were not outright banned, and former members often remained influential in politics and administration. Stephen Holmes of the University of Chicago observed that efforts to purge communist influence were largely unsuccessful, as nearly every family had connections to the former regime and the pressing economic challenges diverted attention from political purges. Exceptions existed, such as in former East Germany, where thousands of former Stasi informers were removed from public positions, but these measures were not replicated uniformly across the region.

In Russia, decommunization efforts were limited and largely symbolic. While the Communist Party of the Soviet Union was banned and replaced with the Communist Party of the Russian Federation, many cities that had been renamed during the Soviet era reverted to their pre-revolutionary names. However, the remnants of Soviet influence persisted, with regions retaining their Soviet-era names despite the changes in major cities. Nostalgia for the Soviet Union has been gradually increasing in Russia, with communist symbols remaining prominent in state-controlled media and political rhetoric. This reverence for Soviet heritage contrasts sharply with the vigorous decommunization efforts in Ukraine, where the post-Soviet government enacted comprehensive laws to remove communist symbols and rename public spaces, actions that were met with staunch opposition from Russia.

Ukraine's decommunization process accelerated following the Revolution of Dignity in 2014, culminating in the 2015 legislation that mandated the removal of communist monuments and renaming streets and cities associated with communist themes. This legislation led to the renaming of over 51,000 streets and the removal of more than a thousand Lenin monuments, underscoring Ukraine's commitment to distancing itself from its Soviet past. The Ukrainian government also took decisive steps to ban communist parties, further severing ties with the ideology that had dominated the region for decades. These measures were part of a broader strategy to forge a new national identity and align more closely with European democratic values.

The legacy of the Cold War continues to influence contemporary international relations. The post-Cold War world is predominantly unipolar, with the United States emerging as the sole superpower. The extensive military alliances and global presence established during the Cold War have left a lasting imprint, as evidenced by the United States maintaining military bases and alliances across the globe. The military-industrial complexes that had flourished during the Cold War have left a legacy of advanced technological infrastructure and substantial military capabilities that continue to shape global power structures.

Economically, the end of the Cold War marked a significant shift. The European Union began expanding eastwards, incorporating many former Warsaw Pact countries and parts of the former Soviet Union, fostering economic integration and promoting democratic governance. However, the transition was uneven, with some nations achieving rapid economic growth and political stability, while others struggled with persistent economic and social challenges. The financial and human costs of the Cold War were immense, with cumulative U.S. military expenditures reaching an estimated $8 trillion and nearly 100,000 American soldiers losing their lives in the Korean and Vietnam Wars. The Soviet Union, too, bore a heavy burden, with its economy strained by extensive military commitments and the loss of life in numerous proxy wars, particularly in Eastern Asia.

Despite the official end of the Cold War, its repercussions continue reverberating. Many regions that were battlegrounds for Cold War tensions, such as the former Yugoslavia, experienced severe civil and ethnic conflicts in the post-Soviet era. The dissolution of state control in various regions led to new conflicts and humanitarian crises, highlighting the enduring instability left in the wake of the Soviet Union's collapse. In contrast, Central and Eastern Europe have embraced mainly economic growth and democratic governance, though other areas, like Afghanistan, have faced ongoing challenges related to state failure and instability.

In summary, the aftermath of the Soviet Union's collapse and the end of the Cold War fundamentally altered the global landscape. The shift from a bipolar to a unipolar world, the economic struggles and political transformations of post-communist states, and the enduring legacy of Cold War institutions and alliances continue to shape international relations today. These changes not only redefined national boundaries and political systems but also set the stage for new forms of cooperation and conflict, influencing everything from economic policies to cultural identities in the decades that followed.

The Second Cold War

The concept of a "New Cold War," often referred to as Cold War II or the Second Cold War, encapsulates the escalating geopolitical tensions that have characterized the early 21st century. Unlike the original Cold War, which primarily pitted the United States against the Soviet Union, the contemporary iteration typically places the United States in opposition to either China or Russia—the latter being the Soviet Union's successor state, which led the Eastern Bloc during the initial Cold War era.

Analysts and commentators have employed various terminology to describe this new phase of global tension. Some use it to draw parallels with the original Cold War, highlighting similarities in ideological confrontations and strategic rivalries. Others caution against this comparison, arguing that the current geopolitical landscape is distinct enough to warrant different terminology. Nevertheless, the notion of a New Cold War often extends beyond bilateral tensions, reflecting the complexities of multilateral relations in today's interconnected world.

Historically, the phrase "new Cold War" emerged as early as the mid-1950s. In 1955, U.S. Secretary of State John Foster Dulles invoked the term, and by 1956, The New York Times warned of Soviet propaganda efforts to rekindle Cold War hostilities. Throughout subsequent decades, scholars and commentators such as Fred Halliday, Alan M. Wald, David S. Painter, and Noam Chomsky have used "Cold War II" interchangeably to describe various phases of the Cold War from 1979 to the early 1990s. In 1975, columnist William Safire suggested that the Nixon administration's policy of détente had failed, signaling the onset of a "Cold War II."

The notion resurfaced in the late 20th century, notably in 1998 when diplomat George Kennan characterized the U.S. Senate's decision to expand NATO to include Poland, Hungary, and the Czech Republic as the commencement of a new Cold War. He anticipated adverse reactions from Russia, which would, in turn, influence its policies adversely. In the early 2000s, experts like James M. Lindsay and Ivo Daalder identified counterterrorism efforts as embodying the "new Cold War," reflecting shifting priorities in global security dynamics.

The term gained renewed traction in the late 2000s and early 2010s. In 2007, Gordon H. Chang referred to the period following President Richard Nixon's historic 1972 meeting with Chinese leader Mao Zedong as "Cold War II." By 2008, British journalist Edward Lucas asserted that a new Cold War between Russia and the West had already begun, a sentiment echoed by various international observers in subsequent years.

The rise of China as a global power has been a significant factor in discussions about a New Cold War. In 2016, the deployment of the Terminal High Altitude Area Defence (THAAD) system in South Korea by the United States sparked concerns in China and Russia, leading to increased mentions of a new Cold War in Chinese media. This development was further compounded by the U.S. support for a tribunal ruling favorable to the Philippines in the South China Sea dispute, exacerbating Sino-American tensions.

Academic discourse in 2019 highlighted differing perspectives on the New Cold War. Steven Lamy and Robert D. English of the University of Southern California debated whether such a conflict was imminent, with English emphasizing the significant threat posed by China in cyberwarfare compared to Russia. Meanwhile, in his 2021 address to the United Nations General Assembly, political leaders like U.S. President Joe Biden explicitly stated that the United States was not seeking a new Cold War, advocating for peaceful coexistence and cooperation on shared global challenges despite existing disagreements.

Niall Ferguson, a senior fellow at the Hoover Institution, posited in 2022 that Cold War II had already commenced, characterized by China's predominant role and Russia's subordinate position. He predicted that the first major conflict of this new era would likely erupt in Europe rather than Asia. This viewpoint was echoed by various international figures and analysts who observed the strengthening alliance between China and Russia, especially in the wake of Russia's invasion of Ukraine in 2022, which further strained relations with the West.

The ongoing rivalry between the United States and China has been a focal point of this New Cold War narrative. During Donald Trump's presidency (2017–2021), rhetoric heightened fears of an emerging Cold War, with Trump frequently labeling China as a significant threat. This period saw increased strategic competition, particularly in technology and trade, though opinions varied on whether these tensions constituted a true Cold War. Scholars like Minxin Pei and commentators such as Michael Collins of the CIA described China's actions as indicative of a "quiet kind of cold war" aimed at supplanting U.S. global dominance.

Under President Joe Biden, the discourse has evolved but remains tense. Biden has consistently denied intentions to ignite a new Cold War, emphasizing the importance of peaceful coexistence and collaboration on global issues. Nevertheless, actions such as the AUKUS security pact and ongoing strategic maneuvers in regions like Africa and the Indo-Pacific continue to fuel debates about the likelihood and nature of a new Cold War.

The term "New Cold War" has also been applied to Russia–United States relations, particularly following Russia's annexation of Crimea in 2014 and its involvement in Eastern Ukraine. These actions have led some analysts to describe the ensuing deterioration in relations as a new Cold War, although this characterization remains contested.

The concept of a "New Cold War," often called Cold War II or the Second Cold War, encapsulates the escalating geopolitical tensions that have characterized the early 21st century. Unlike the original Cold War, which primarily pitted the United States against the Soviet Union, the contemporary iteration typically places the United States in opposition to either China or Russia—the latter being the Soviet Union's successor state, which led the Eastern Bloc during the initial Cold War era.

Analysts and commentators have employed various terminology to describe this new phase of global tension. Some use it to draw parallels with the original Cold War, highlighting similarities in ideological confrontations and strategic rivalries. Others caution against this comparison, arguing that the current geopolitical landscape is distinct enough to warrant different terminology. Nevertheless, the notion of a New Cold War often extends beyond bilateral tensions, reflecting the complexities of multilateral relations in today's interconnected world.

Historically, the phrase "new Cold War" emerged as early as the mid-1950s. In 1955, U.S. Secretary of State John Foster Dulles invoked the term, and by 1956, The New York Times warned of Soviet propaganda efforts to rekindle Cold War hostilities. Throughout subsequent decades, scholars and commentators such as Fred Halliday, Alan M. Wald, David S. Painter, and Noam Chomsky have used "Cold War II" interchangeably to describe various phases of the Cold War from 1979 to the early 1990s. In 1975, columnist William Safire suggested that the Nixon administration's policy of détente had failed, signaling the onset of a "Cold War II."

The notion resurfaced in the late 20th century, notably in 1998 when diplomat George Kennan characterized the U.S. Senate's decision to expand NATO to include Poland, Hungary, and the Czech Republic as the commencement of a new Cold War. He anticipated adverse reactions from Russia, which would, in turn, influence its policies adversely. In the early 2000s, experts like James M. Lindsay and Ivo Daalder identified counterterrorism efforts as embodying the "new Cold War," reflecting shifting priorities in global security dynamics.

The term gained renewed traction in the late 2000s and early 2010s. In 2007, Gordon H. Chang referred to the period following President Richard Nixon's historic 1972 meeting with Chinese leader Mao Zedong as "Cold War II." By 2008, British journalist Edward Lucas asserted that a new Cold War between Russia and the West had already begun, a sentiment echoed by various international observers in subsequent years.

The rise of China as a global power has been a significant factor in discussions about a New Cold War. In 2013, Michael Klare compared tensions between Russia and the West to the ongoing proxy conflict between Saudi Arabia and Iran, highlighting the multifaceted nature of contemporary geopolitical struggles. Oxford Professor Philip N. Howard argued that a new cold war was being fought via the media, information warfare, and cyberwar, underscoring the evolving tactics of modern statecraft.

In 2016, the deployment of the Terminal High Altitude Area Defence (THAAD) system in South Korea by the United States sparked concerns in China and Russia, leading to increased mentions of a new Cold War in Chinese media. This development was further compounded by the U.S. support for a tribunal ruling favorable to the Philippines in the South China Sea dispute, exacerbating Sino-American tensions. Concurrently, senior UK government officials expressed growing fears of a new Cold War unfolding in Europe, citing Russian efforts to undermine European unity on strategic issues.

The Syrian Civil War emerged as another flashpoint, with observers, including Syrian President Bashar al-Assad, viewing the conflict as a proxy war between Russia and the United States, and some even labeling it a "proto-world war." In April 2018, relations deteriorated further over potential U.S.-led military strikes in the Middle East following the Douma chemical attack in Syria and the poisoning of former Russian spy Sergei Skripal in the UK. United Nations Secretary-General António Guterres remarked that the Cold War had returned with greater intensity, noting the absence of safeguards to manage such crises effectively.

The escalation continued with Russia's 2014 annexation of Crimea, which many political analysts argue marked the beginning of a new Cold War between Russia and the West. By August 2014, both sides had imposed economic, financial, and diplomatic sanctions upon each other, with Western countries led by the U.S. and European Union implementing punitive measures and Russia introducing retaliatory actions. Prominent figures like Mikhail Gorbachev warned that the world was on the brink of a new Cold War amid the Russo-Ukrainian conflict, though some scholars contended that the term did not accurately capture the nature of contemporary relations between Russia and the West.

The Russian invasion of Ukraine in February 2022 significantly intensified these tensions. The full-scale invasion led to widespread international condemnation and further solidified the perception of a new Cold War. Journalists and historians debated the implications, with some, like Gideon Rachman, asserting that the invasion marked the definitive start of a second Cold War, while others, such as Arne Westad and Fredrik Logevall, argued that the situation did not fit the traditional Cold War framework.

In the evolving global landscape, the rivalry between the United States and China has emerged as a defining aspect of modern geopolitical tensions, often referred to as a "New Cold War." During Donald Trump's presidency from 2017 to 2021, this rivalry intensified as Trump's rhetoric frequently framed China as a significant threat. Strategic competition between the two nations escalated, particularly in areas of technology and trade. While the heightened tensions drew comparisons to the Cold War, scholars like Minxin Pei and commentators such as Michael Collins from the CIA described China's actions as indicative of a "quiet cold war" aimed at challenging U.S. global dominance. The debate over whether these tensions constituted a true Cold War was a matter of ongoing discourse.

The methods of competition in this New Cold War have shifted from the overt military and ideological confrontations of the 20th century to more subtle, yet equally potent, tools such as media manipulation and cyber warfare. As Oxford Professor Philip N. Howard emphasized, the battle between global powers is increasingly being fought through information warfare and cyber operations. These modern tactics reflect the changing face of geopolitical competition, where control over narratives and digital landscapes can be as impactful as control over physical territories.

Even as the focus on U.S.-China relations grew, concerns about Russia's actions remained central to the New Cold War narrative. In early 2016, UK government officials voiced apprehensions about Russia's attempts to destabilize European unity, further reinforcing the idea that a new form of Cold War was emerging across multiple fronts. This notion was exacerbated by Russia's annexation of Crimea in 2014 and its involvement in Eastern Ukraine, which led many to describe the ensuing deterioration in U.S.-Russia relations as a modern Cold War. Analysts like former MI6 chief John Sawers warned that the current geopolitical climate could be more dangerous than the original Cold War due to the absence of established strategic relationships and the unpredictability of modern state actors.

Under President Joe Biden, who took office in 2021, the discourse surrounding this new era of geopolitical rivalry has evolved, yet the tensions persist. In his address to the United Nations General Assembly in September 2021, Biden explicitly stated that the United States was not seeking a new Cold War, advocating instead for peaceful coexistence and cooperation on shared global challenges. However, despite this diplomatic rhetoric, actions such as the AUKUS security pact—a defense agreement between the U.S., UK, and Australia—and ongoing strategic maneuvers in regions like Africa and the Indo-Pacific have continued to fuel debates about the true nature of the United States' competition with China and Russia.

The academic discourse surrounding this New Cold War has been complex and multi-faceted. In 2019, scholars like Steven Lamy and Robert D. English from the University of Southern California debated whether a new Cold War was imminent, with English focusing on the significant cyber threats posed by China in comparison to Russia. While political leaders like Biden have emphasized the importance of avoiding direct confrontation, others, such as Niall Ferguson, a senior fellow at the Hoover Institution, have argued that Cold War II is already underway. Ferguson's view, expressed in 2022, characterized this new global conflict as being driven by China's dominant role and Russia's more subordinate position. He even predicted that the first major conflict of this new Cold War would likely occur in Europe, a forecast that appeared prescient as Russia's invasion of Ukraine in 2022 further strained relations between Russia and the West, while also solidifying the emerging alliance between China and Russia.

The dynamics of this New Cold War are not confined to bilateral rivalries. The global landscape is shaped by shifting alliances and the involvement of multiple nations, making the situation more intricate than the binary opposition seen during the original Cold War. For example, the Syrian Civil War has been viewed by some as a proxy battleground between Russia and the United States, further entangling regional conflicts within the broader context of global power struggles.

In addition to military and strategic competition, the role of information warfare has transformed the nature of geopolitical conflict. Cyberattacks and media manipulation have become critical tools in this new era, as nations seek to undermine each other's influence without direct confrontation. This multifaceted approach to conflict underscores the complexity of modern international relations, where warfare can be waged across multiple fronts—diplomatic, economic, technological, and informational.

As the global landscape continues to evolve, the concept of a New Cold War remains a potent framework for understanding the intricate and often contentious interactions that define international relations in the 21st century. While the historical parallels with the original Cold War are clear, the unique characteristics of today's geopolitical environment—marked by economic interdependence, rapid technological advancements, and complex multilateral relationships—reveal both similarities and significant differences from the past. The rivalry between the United States, China, and Russia, combined with strategic maneuvering in various regions, has created a dynamic and often unpredictable global environment. This ongoing struggle for dominance reflects the realities of contemporary global politics and shapes the narrative of a New Cold War, where the stakes are high, and the outcomes remain uncertain.

Chapter 3 - The 1947 National Security Act

From the early 1920s through the mid-1940s, the United States grappled with the formidable task of reorganizing its armed forces better to meet the demands of a rapidly changing global landscape. During this period, Congress considered approximately fifty bills to restructure the military. However, persistent opposition from both the Department of the Navy and the War Department stymied these efforts. Nearly all proposed legislation failed to advance beyond the House floor, and the sole bill that did reach a vote was decisively defeated in 1932 by a margin of 153 to 135.

The outbreak of World War II fundamentally altered the considerations surrounding military organization. The complexities and exigencies of a global conflict underscored the need for a more unified and efficient military structure. Several critical factors had converged by the war's end, compelling American leaders to seriously contemplate comprehensive military reform to enhance unity and operational effectiveness.

This momentum culminated in the National Security Act of 1947, enacted on July 26, 1947 (Public Law 80-253, 61 Stat. 495). The legislation marked a pivotal restructuring of the United States government's military and intelligence apparatus in the aftermath of World War II. Most provisions of the Act became effective on September 18, 1947, immediately following the Senate's confirmation of James Forrestal as the first Secretary of Defense.

The National Security Act of 1947 orchestrated the Department of the Army merger, formerly known as the Department of War, the Department of the Navy, and the newly established Department of the Air Force into a single entity called the National Military Establishment (NME). This unification aimed to foster greater coordination and eliminate inter-service rivalries that had previously hampered military effectiveness. The Act also established the position of Secretary of Defense, tasked with overseeing the NME and ensuring cohesive military strategy and administration.

A significant outcome of the Act was the formal establishment of the United States Air Force as an independent branch, distinct from the Army Air Forces. This move recognized the strategic importance of air power in modern warfare and aimed to provide the Air Force with the autonomy necessary to develop its capabilities. Additionally, the Act maintained the Marine Corps as an independent service under the Department of the Navy, ensuring its specialized role within the broader military framework.

Beyond restructuring the military departments, the National Security Act of 1947 laid the groundwork for the United States' intelligence community. It established the National Security Council (NSC), designed to advise the President on national security and foreign policy matters, thereby integrating military and civilian oversight. The Act also created the Central Intelligence Agency (CIA), headed by the Director of Central Intelligence, to coordinate intelligence activities and provide strategic analysis to policymakers.

The passage of the National Security Act was the culmination of persistent efforts by President Harry S. Truman, who had advocated for comprehensive military and intelligence reform since 1944. Truman formally proposed the legislation to Congress on February 26, 1947, recognizing the imperative need for a unified defense structure in the face of emerging global threats. The bill was swiftly introduced in the House of Representatives on February 28, 1947, and in the Senate on March 3, 1947, with Senator Chan Gurney serving as its sponsor.

As chairman of the Senate Committee on Armed Services, Senator Gurney spearheaded committee hearings from mid-March to early May, navigating the bill through intense debates and overcoming significant resistance. The legislation garnered bipartisan support, reflecting a broad consensus on the necessity of restructuring the nation's defense framework. The Senate passed the bill on July 9, 1947, followed by the House on July 19, 1947. Additionally, the Senate approved a related House resolution (80 H.Con.Res. 70) on July 16, 1947, further solidifying the Act's legislative foundation.

President Truman signed the National Security Act into law aboard his VC-54C presidential aircraft, the Sacred Cow, on July 26, 1947. This symbolic gesture underscored the legislation's transformative nature and significance for the future of the United States' military and intelligence operations.

The National Security Act of 1947 fundamentally reshaped the American defense establishment, promoting greater unity and coordination among the military branches while establishing critical institutions that continue to influence national security policy to this day. By addressing the shortcomings revealed during World War II, the Act laid the foundation for a more integrated and responsive military apparatus, capable of addressing the complexities of both conventional and unconventional threats in the ensuing Cold War era and beyond.

The National Security Agency

The National Security Agency (NSA) stands as a cornerstone of the United States' intelligence apparatus, operating under the auspices of the Department of Defense and overseen by the Director of National Intelligence (DNI). Tasked with the global monitoring, collection, and processing of information and data, the NSA specializes in signals intelligence (SIGINT), a discipline focused on intercepting and analyzing electronic communications. Beyond intelligence gathering, the NSA is also responsible for safeguarding U.S. communications networks and information systems, employing a range of clandestine measures to fulfill its mission. With an estimated workforce of approximately 32,000 employees, the NSA plays a pivotal role in national security.

The origins of the NSA can be traced back to World War II, where it began as a unit dedicated to deciphering coded communications. Officially established by President Harry S. Truman in 1952, the agency rapidly expanded during the Cold War, becoming the largest U.S. intelligence organization in terms of personnel and budget. However, by 2013, the Central Intelligence Agency (CIA) had surpassed the NSA in budgetary terms, reflecting shifts in intelligence priorities and operational focuses.

In the contemporary era, the NSA conducts extensive worldwide data collection, utilizing both sophisticated cyber surveillance and physical eavesdropping techniques. One notable method has been the deployment of electronic bugs to infiltrate and monitor critical systems, a strategy exemplified by the alleged creation of the Stuxnet virus, which significantly disrupted Iran's nuclear program. The NSA, in collaboration with the CIA, operates the highly classified Special Collection Service (SCS), which embeds eavesdropping devices in high-value targets such as presidential palaces and embassies. The SCS employs a variety of tactics, including close surveillance, burglary, wiretapping, and covert entry, to gather intelligence.

Distinct from the CIA and the Defense Intelligence Agency (DIA), which primarily focus on foreign human intelligence (HUMINT), the NSA does not publicly engage in human intelligence gathering. Instead, it provides critical SIGINT support to other government agencies, ensuring that intelligence activities remain coordinated and effective. The Central Security Service (CSS), a co-located organization within the NSA, facilitates cooperation between the agency and other U.S. defense cryptanalysis units. To streamline communication and enhance operational efficiency, the Director of the NSA concurrently serves as the Commander of the United States Cyber Command and the Chief of the Central Security Service.

The NSA's operations have not been without controversy. The agency has faced significant political scrutiny for its surveillance practices, including the interception of communications of anti–Vietnam War leaders and involvement in economic espionage. The most notable public revelation came in 2013, when Edward Snowden, a former NSA contractor, disclosed extensive details about the agency's secret surveillance programs. These disclosures revealed that the NSA was intercepting and storing the communications of over a billion individuals worldwide, including American citizens, and tracking the movements of hundreds of millions through cellphone metadata. Internationally, research has demonstrated the NSA's capability to surveil domestic internet traffic of foreign nations through techniques such as "boomerang routing."

The history of the NSA is marked by continuous evolution in response to emerging threats and technological advancements. Its formation in 1952 marked the beginning of a new era in intelligence, one that emphasized electronic surveillance and cyber capabilities. Over the decades, the NSA has adapted to the changing landscape of global security, expanding its reach and refining its methods to address both conventional and unconventional threats.

The establishment of the NSA was part of a broader effort to create a unified and efficient intelligence community. The National Security Act of 1947 had already set the stage by reorganizing the military and intelligence agencies, leading to the creation of the National Security Council and the Central Intelligence Agency. The NSA's formation built upon this foundation, integrating advanced cryptographic and surveillance technologies to enhance national security.

In the post-Cold War era, the NSA has continued to play a crucial role in countering global threats such as terrorism, cyberattacks, and the proliferation of weapons of mass destruction. Its ability to intercept and analyze vast amounts of data has made it an indispensable tool in the United States' intelligence and defense strategies. However, the balance between security and privacy remains a contentious issue, with ongoing debates about the appropriate scope and oversight of the agency's activities.

Today, the NSA remains at the forefront of the United States' efforts to protect national security in an increasingly digital and interconnected world. Its capabilities in signals intelligence and cyber defense are continually evolving, ensuring that the agency remains a key player in both national and global security arenas. As technology advances and new threats emerge, the NSA's role and methods will undoubtedly continue to adapt, reflecting the dynamic nature of modern intelligence work.

In summary, the National Security Agency is a critical component of the United States' defense and intelligence infrastructure, specializing in signals intelligence and cyber defense. From its origins in World War II to its current status as a leading intelligence agency, the NSA has continuously evolved to meet the demands of national security. While its extensive surveillance capabilities have generated significant political controversy, the NSA remains indispensable in the ongoing effort to protect the United States from a wide array of global threats.

The US Air Force

The United States Air Force (USAF), one of the eight uniformed services of the United States, serves as the air service branch of the United States Armed Forces. Established on August 1, 1907, originally as part of the United States Army Signal Corps, the USAF became a separate branch in 1947 through the National Security Act of that year. As the second youngest branch of the Armed Forces and fourth in order of precedence, the Air Force has developed a comprehensive set of core missions that include air supremacy, global integrated intelligence, surveillance and reconnaissance (ISR), rapid global mobility, global strike, and command and control.

Organized within the Department of the Air Force, one of the three military departments under the Department of Defense, the USAF is led by the civilian Secretary of the Air Force, who reports directly to the Secretary of Defense. Appointed by the President and confirmed by the Senate, the Secretary oversees the Air Force's administrative functions. The highest-ranking military officer within the Air Force is the Chief of Staff of the Air Force, a member of the Joint Chiefs of Staff, who supervises Air Force units and ensures operational readiness. Under the direction of the Secretary of Defense and the Secretary of the Air Force, specific Air Force components are assigned to unified combatant commands, where combatant commanders are granted operational authority over these forces while the Secretary and Chief of Staff retain administrative control.

The USAF plays a multifaceted role in military operations, conducting independent air missions while also providing essential air support to land and naval forces. This support extends to the recovery of troops in the field, demonstrating the Air Force's versatility and integral position within the broader military framework. As of 2020, the Air Force operates approximately 5,500 military aircraft and maintains around 400 intercontinental ballistic missiles (ICBMs), making it the world's largest air force. With a budget of $179.7 billion, it stands as the second-largest service branch in the U.S. Armed Forces, supported by a workforce of over 321,000 active duty airmen, nearly 148,000 civilian personnel, approximately 69,000 reserve airmen, over 105,000 Air National Guard members, and around 65,000 Civil Air Patrol auxiliaries.

The foundational missions of the USAF, as outlined by the National Security Act of 1947, emphasize the preparation and maintenance of air forces for effective warfare, both offensive and defensive. These missions have evolved over time but remain focused on ensuring global vigilance, reach, and power. Air superiority, one of the core missions, involves achieving dominance in the air to allow for unhindered operations by friendly forces. This encompasses both Offensive Counter-Air (OCA) operations, which seek to destroy or neutralize enemy air capabilities near their source, and Defensive Counter-Air (DCA) measures, which protect friendly airspace from enemy incursions through active and passive defenses.

Global integrated ISR represents the synchronization of sensors, assets, and information systems worldwide to support current and future operations. This mission involves the meticulous planning and execution of intelligence collection, processing, analysis, and dissemination to provide decision-makers with accurate and timely information. Rapid global mobility, another cornerstone mission, ensures the timely deployment and sustainment of military forces and capabilities across the globe. This is achieved through airlift operations, aerial refueling, and aeromedical evacuation, enabling the Air Force to project power swiftly and respond to crises with agility and precision.

The global strike mission underscores the Air Force's ability to conduct precise and sustained attacks against strategic targets, leveraging a diverse array of munitions to achieve swift and decisive effects. This mission includes strategic bombing, air interdiction, and close air support, all of which are critical for weakening adversaries and supporting ground operations. Central to this mission is nuclear deterrence, which involves maintaining and securing nuclear forces to prevent adversaries from taking aggressive actions against U.S. interests. The establishment of the Air Force Global Strike Command in 2008 marked a significant enhancement of the Air Force's nuclear capabilities, ensuring a credible and effective deterrent.

Command and control (C2) functions are integral to the Air Force's operations, encompassing the exercise of authority and direction by designated commanders over assigned forces. Effective C2 ensures the coordination and integration of personnel, equipment, communications, and procedures to achieve strategic, operational, and tactical objectives. At the strategic level, C2 involves setting national security objectives and guiding the allocation of resources to meet these goals. Operational C2 focuses on planning and executing campaigns and major operations within specific theaters, while tactical C2 manages individual battles and engagements to achieve immediate objectives.

Throughout its history, the USAF has been instrumental in numerous conflicts and operations, from its early days in the Mexican Expedition and both World Wars to the Cold War, Korean War, Vietnam War, and contemporary engagements in the Middle East and Africa. The Air Force's extensive involvement in these operations highlights its critical role in U.S. military strategy and its evolution into a technologically advanced and strategically vital force.

In the 21st century, the USAF has continued to adapt to emerging threats and technological advancements. Early in the 2000s, procurement projects such as the KC-X tanker and the F-35 fighter program faced delays, leading to an increase in the average age of aircraft within the fleet. Concurrently, the Air Force has placed a strong emphasis on improving Basic Military Training (BMT) for enlisted personnel, incorporating more comprehensive deployment simulations to better prepare trainees for real-world combat environments. Budget constraints in 2007 prompted a Reduction-in-Force (RIF), aiming to decrease active duty personnel from 360,000 to 316,000. However, this reduction was partially reversed in 2008 to meet operational demands, resulting in a maintained force size of approximately 330,000 personnel.

A significant turning point occurred on June 5, 2008, when Secretary of Defense Robert Gates accepted the resignations of Secretary of the Air Force Michael Wynne and Chief of Staff General T. Michael Moseley. Citing systemic issues related to the Air Force's nuclear mission focus and performance, Gates underscored the need to reinforce the USAF's nuclear capabilities. This led to the creation of the Air Force Global Strike Command, which took control of all bomber aircraft, thereby reinforcing the Air Force's commitment to nuclear deterrence.

In subsequent years, the USAF continued to refine its force structure and strategic priorities. In 2009, the Air Force released a plan that reduced the number of fighter aircraft while reallocating resources to support nuclear, irregular, and information warfare. The same year, the Unmanned Aerial System (UAS) Flight Plan was introduced, outlining ambitious goals for unmanned and hypersonic technologies extending to 2047. Predictions by Air Force Chief Scientist Greg Zacharias included the development of hypersonic weapons in the 2020s, hypersonic unmanned aerial vehicles (RPAs) by the 2030s, and recoverable hypersonic RPAs by the 2040s, alongside plans to deploy a sixth-generation jet fighter by the mid-2030s.

World War II was a defining period for the Air Forces, with nearly 68,000 airmen losing their lives in service—a number surpassed only by infantry casualties. The war underscored the critical role of air supremacy, intelligence, surveillance, and rapid global mobility in modern warfare. Despite functioning autonomously, airmen continued to advocate for formal independence, a goal realized with the National Security Act of 1947. Signed on July 26, 1947, this legislation established the Department of the Air Force as a separate branch within the newly formed National Military Establishment, later renamed the Department of Defense in 1949. The Air Force officially emerged as an independent service branch on September 18, 1947, when W. Stuart Symington was sworn in as its first Secretary.

The post-war era marked significant advancements and restructuring within the USAF. In 1947, Captain Chuck Yeager achieved a monumental milestone by breaking the sound barrier in the X-1 rocket-powered aircraft, ushering in a new era of aeronautics. Throughout the Cold War, the Air Force played a pivotal role in maintaining air supremacy and developing advanced aircraft such as the SR-71 Blackbird and the F-117 Nighthawk, the latter of which was retired from combat service in April 2008.

Entering the 21st century, the USAF faced new challenges and transformations. Early in the 2000s, procurement projects like the KC-X tanker and the F-35 fighter program experienced delays, resulting in the Air Force setting new records for average aircraft age. Concurrently, the USAF emphasized improving Basic Military Training (BMT) for enlisted personnel. Since 2005, training programs evolved to include the BEAST deployment phase, simulating combat environments to better prepare trainees for real-world deployments. In November 2022, the BEAST program was replaced by PACER FORGE, reflecting ongoing efforts to enhance training effectiveness.

Budget constraints led to significant changes within the Air Force in 2007, initiating a Reduction-in-Force (RIF) that aimed to decrease active duty personnel from 360,000 to 316,000. However, this reduction was curtailed in 2008, maintaining personnel levels around 330,000 to meet the demands of combatant commanders and mission requirements. These financial limitations also resulted in a sharp decline in flight hours for crew training and directed the Deputy Chief of Staff for Manpower and Personnel to implement Airmen's Time Assessments.

A pivotal moment occurred on June 5, 2008, when Secretary of Defense Robert Gates accepted the resignations of both the Secretary of the Air Force, Michael Wynne, and the Chief of Staff of the Air Force, General T. Michael Moseley. Citing systemic issues related to the declining focus and performance of the Air Force's nuclear mission, Gates emphasized the need to reinforce the USAF's nuclear capabilities. This led to the establishment of the Air

Force Global Strike Command on October 24, 2008, which assumed control of all bomber aircraft, underscoring the Air Force's commitment to maintaining nuclear deterrence.

In 2009, the USAF unveiled a force structure plan that prioritized resources toward nuclear, irregular, and information warfare while reducing the number of fighter aircraft. Additionally, the Unmanned Aerial System (UAS) Flight Plan was released, outlining ambitious plans for unmanned and hypersonic technologies through 2047. Air Force Chief Scientist Greg Zacharias projected the development of hypersonic weapons in the 2020s, hypersonic unmanned aerial vehicles (RPAs) by the 2030s, and recoverable hypersonic RPAs by the 2040s. The USAF also announced intentions to deploy a sixth-generation jet fighter by the mid-2030s, reflecting its ongoing commitment to technological advancement.

The USAF continued to engage in international collaborations and operations into the 2020s. On October 22, 2023, the Air Force conducted its first-ever trilateral exercise with South Korean and Japanese air forces near the Korean Peninsula, highlighting its role in regional security. However, challenges persisted, as evidenced by the tragic crash of a Bell Boeing V-22 Osprey in Yakushima, Japan, on November 29, 2023, resulting in the loss of one airman.

In 2024, the Air Force faced environmental and legal challenges, notably refusing to comply with an Environmental Protection Agency (EPA) order to develop a cleanup plan for drinking water around Tucson, Arizona. This decision followed the Supreme Court's ruling in Loper Bright Enterprises v. Raimondo, amid concerns over PFAS contamination from nearby Air Force bases.

Throughout its history, the USAF has been integral to numerous conflicts and operations, from the Mexican Expedition and both World Wars to the Cold War, Korean War, Vietnam War, and contemporary conflicts in the Middle East and Africa. The Air Force's extensive involvement in these operations underscores its vital role in U.S. military strategy and its evolution into a modern, technologically advanced force.

As the USAF continues to adapt to emerging threats and technological advancements, its legacy of innovation and strategic prowess remains a cornerstone of American defense. From the early days of balloon reconnaissance to the cutting-edge developments of the 21st century, the United States Air Force exemplifies the enduring pursuit of air supremacy and the relentless drive to maintain a strategic advantage in an ever-changing global landscape.

The Central Intelligence Agency

The Central Intelligence Agency (CIA) plays a vital, though often understated, role in supporting the United States Space Force, primarily through intelligence gathering, analysis, and strategic cooperation. While the Space Force is a military branch focused on safeguarding U.S. interests in space, the CIA contributes by providing critical intelligence on foreign space capabilities, emerging technologies, and potential threats in the increasingly contested space domain.

The CIA's role in this collaboration centers on the collection and analysis of foreign space operations and developments, particularly those of adversarial nations like China and Russia, whose advancements in satellite technology, anti-satellite weapons, and space-based surveillance pose strategic challenges to U.S. space security. Through its Directorate of Science and Technology and Directorate of Operations, the CIA monitors and assesses these capabilities, ensuring that the Space Force has actionable intelligence to shape its strategic priorities and defense initiatives in space.

Another key aspect of the CIA's involvement is its coordination with the Space Force on counterintelligence efforts. As space technology is a critical area of both military and economic competition, protecting U.S. assets from espionage and technological theft is a top priority. The CIA helps identify and mitigate espionage activities that target American space infrastructure and innovation, providing a critical layer of security in the broader U.S. space defense strategy.

Additionally, the CIA plays a strategic role in monitoring the global space landscape, offering insight into how other nations are militarizing space, investing in space exploration, or developing space-based weapons systems. This intelligence allows the Space Force to anticipate future threats and align its mission to maintain U.S. dominance in space operations.

The Central Intelligence Agency (CIA), often referred to simply as the Agency, serves as the civilian foreign intelligence service of the United States, responsible for gathering, processing, and analyzing national security information from around the world. Primarily using human intelligence (HUMINT), the CIA is crucial in providing intelligence for the president, the Cabinet, and senior policymakers. Founded in the aftermath of World War II, the CIA emerged from the Office of Strategic Services (OSS), dissolved by President Harry S. Truman. In 1946, Truman established the Central Intelligence Group, which later evolved into the CIA with the passage of the National Security Act of 1947. This new agency was tasked with consolidating intelligence gathering and covert operations under a central authority.

Unlike domestic agencies such as the Federal Bureau of Investigation (FBI), the CIA focuses on foreign intelligence with a limited domestic role. It serves as the national manager for HUMINT, coordinating intelligence operations across the broader U.S. Intelligence Community (IC). Additionally, the agency is empowered to conduct covert actions on behalf of the president, often through its Directorate of Operations, which oversees clandestine activities.

Throughout its history, the CIA has exerted significant political influence abroad. Its paramilitary units, particularly those within the Special Activities Center, have been instrumental in covert interventions, including regime changes and support for foreign political groups. The agency has a long history of establishing intelligence services in allied nations, such as Germany's Federal Intelligence Service, and has provided operational support in areas ranging from training to technical assistance.

Since 2004, the CIA has operated under the Office of the Director of National Intelligence (ODNI), which was created in response to the intelligence failures that led to the September 11 attacks. Despite the shift in oversight, the CIA has continued to grow, especially in areas related to counterterrorism and offensive cyber operations. By 2010, it had the largest budget of any intelligence agency, a reflection of its expanded role in global operations and national security. One of the largest divisions within the CIA, the Information Operations Center, has evolved from a counterterrorism focus to offensive cyber operations, signaling the agency's adaptation to new forms of warfare.

The CIA's activities have not been without controversy. Over the years, the agency has been implicated in numerous scandals, including its use of torture, unauthorized surveillance, and covert operations that have drawn criticism both domestically and internationally. In 2022, revelations of a domestic surveillance program underscored ongoing concerns about the agency's transparency and accountability. Despite these challenges, the CIA remains a central pillar of U.S. national security.

Upon its founding, the CIA was intended to serve as a clearinghouse for foreign policy intelligence. Its mandate was to collect, analyze, and disseminate foreign intelligence, while also conducting covert operations. By 2013, the agency's core priorities included counterterrorism, nonproliferation of weapons of mass destruction, counterintelligence, and cyber intelligence—reflecting the evolving global security landscape.

The CIA's organizational structure consists of five key directorates: Digital Innovation, Analysis, Operations, Science and Technology, and Support. The Directorate of Digital Innovation (DDI), the newest of these, was formally established in 2015 to integrate digital and cybersecurity capabilities into the agency's espionage and covert operations. It plays a pivotal role in cyber operations, equipping CIA officers with the tools needed to engage in offensive cyber activities. Although little is publicly known about the directorate's specific operations, its creation highlights the CIA's adaptation to the digital age.

The Directorate of Analysis, historically known as the Directorate of Intelligence, is tasked with synthesizing intelligence for the president and senior leaders, providing them with critical insights into global security issues. The directorate is divided into regional and issue-based groups that focus on key areas such as the Near East, Russia, and counterterrorism, ensuring a broad and nuanced understanding of global threats.

The Directorate of Operations oversees the agency's clandestine activities, primarily HUMINT operations. It is responsible for coordinating intelligence collection across the U.S. Intelligence Community. Despite efforts to streamline operations between the CIA and the Department of Defense (DOD), rivalry between the two organizations has persisted, particularly with the DOD's establishment of its own clandestine service, the Defense Clandestine Service (DCS), under the Defense Intelligence Agency (DIA).

Meanwhile, the Directorate of Science and Technology focuses on the development and deployment of advanced technologies for intelligence collection. Among its many innovations was the development of the U-2 reconnaissance aircraft, which played a key role in gathering intelligence over Soviet territory during the Cold War. The CIA's technological achievements have often been shared with other agencies, including the military, further blurring the lines between intelligence gathering and defense operations.

The Directorate of Support provides logistical and operational support to the entire agency, managing key functions such as communications, security, and information technology. Its contributions are vital to the smooth functioning of the CIA's global operations.

The agency's leadership consists of the Director of the CIA (D/CIA), who is appointed by the president and confirmed by the Senate. The director reports to the director of national intelligence (DNI) but also interfaces directly with the White House and Congress. The deputy director (DD/CIA), who manages the CIA's day-to-day operations, is appointed by the director without Senate confirmation, making it one of the highest-ranking non-political positions within the agency.

Throughout its existence, the CIA has evolved into a multifaceted organization, adapting to new global threats and technological advancements. From its early days focused on Cold War espionage to its current role in cyber warfare and counterterrorism, the agency continues to be a cornerstone of U.S. national security efforts. Despite its controversies, the CIA's influence on both international relations and the internal security of the United States remains profound.

The success of British Commandos during World War II inspired U.S. President Franklin D. Roosevelt to establish a comparable intelligence service modeled after the British Secret Intelligence Service (MI6) and the Special Operations Executive. On June 13, 1942, Roosevelt issued a presidential military order that created the Office of Strategic Services (OSS). General William J. Donovan, who had served as a key advocate for a centralized intelligence organization, led the OSS. Donovan envisioned a global intelligence service that could counter threats, particularly communism, and provide vital information directly to the president. His proposal for a "Central Intelligence Service" in 1944 aimed to continue the work of the OSS in peacetime, marking the origins of what would eventually become the CIA.

Following the death of Roosevelt in 1945, President Harry Truman inherited a presidency with limited insight into global intelligence operations and key wartime projects. Unlike Donovan's vision of a comprehensive intelligence organization, Truman initially imagined the new intelligence body as a global information-gathering service, more akin to a news organization than a clandestine spy network. His concern was to avoid creating a U.S. equivalent of the Gestapo, the notorious secret police of Nazi Germany.

On September 20, 1945, shortly after World War II ended, Truman signed an executive order dissolving the OSS. Its functions were divided between the State Department and the War Department. However, this arrangement proved short-lived. In late 1945, the idea of a "Central Intelligence Agency" began to take shape in discussions among key figures, including Secretary of the Navy James Forrestal and Commander Arthur Radford, who presented a restructuring proposal to the U.S. Senate Military Affairs Committee. Colonel Sidney Mashbir and Commander Ellis Zacharias, under the direction of Fleet Admiral Joseph King, collaborated on a draft to establish what would eventually become the CIA.

Despite initial opposition from various corners of the U.S. government—including the military, State Department, and FBI—President Truman moved forward with the creation of a central intelligence body. In January 1946, he established the National Intelligence Authority (NIA), with its operational branch known as the Central Intelligence Group (CIG), the direct predecessor of the CIA. The CIG marked the beginning of a more formalized intelligence service, but the CIA would not officially come into existence until the passage of the National Security Act of 1947, which Truman signed into law on July 26 of that year.

The creation of the CIA was driven by escalating tensions with the Soviet Union in the early days of the Cold War. Lawrence Houston, legal counsel for the Strategic Services Unit (SSU), the CIG, and later the CIA, was instrumental in drafting the National Security Act of 1947. This legislation not only established the CIA but also created the National Security Council (NSC) to coordinate national security policy. Two years later, Houston helped draft the Central Intelligence Agency Act of 1949, which granted the CIA extensive financial and operational flexibility, allowing it to operate with a high level of secrecy.

At the onset of the Korean War, the CIA was still in its infancy, with only a few thousand employees, of which about a thousand worked in analysis. Intelligence was primarily gathered from State Department telegrams, military dispatches, and public documents, as the agency had yet to develop its intelligence-gathering capabilities fully. The lack of a clear warning about the North Korean invasion of South Korea in 1950 was seen as a significant intelligence failure, leading Truman to appoint Walter Bedell Smith as the new Director of the CIA. Smith's appointment came at a critical juncture, as the agency faced increasing demands for intelligence in a rapidly intensifying Cold War environment.

The CIA's evolving role required it to balance multiple demands from the president, the Department of Defense, and the State Department. Truman sought a centralized intelligence organization that could streamline the flow of information to his desk, while the Department of Defense pushed for military intelligence and covert action capabilities. At the same time, the State Department focused on leveraging intelligence to effect global political change favorable to U.S. interests. This led to the CIA's dual mission of conducting both covert intelligence operations and covert action.

U.S. Air Force General Hoyt Vandenberg, the second director of the CIG, established the Office of Special Operations (OSO) and the Office of Reports and Estimates (ORE), which became central to the agency's espionage and analytical activities. Vandenberg focused squarely on gathering intelligence about Soviet military movements and capabilities in Eastern and Central Europe. In 1948, the National Security Council issued Directive 10/2, which authorized covert operations against hostile foreign powers, particularly the Soviet Union. To carry out these operations, the CIA created the Office of Policy Coordination (OPC), which conducted secretive actions even outside the direct oversight of the CIA director, reporting instead to the secretaries of defense and state, as well as the NSC.

Despite its growing influence, the CIA suffered several early intelligence failures, including its inability to predict the Soviet Union's acquisition of nuclear weapons, the Soviet blockade of Berlin, and the Chinese entry into the Korean War. Moreover, the CIA's operations were compromised by Soviet double agents such as Kim Philby, who, as a British liaison, was privy to many of the CIA's clandestine plans in Eastern Europe, leading to the failure of numerous missions behind the Iron Curtain.

Nevertheless, the CIA did have some early successes. One of its most notable operations was its influence on the 1948 Italian general election, where the agency funneled funds to the Christian Democrats to prevent the Communist Party from gaining power. The CIA used the U.S. Exchange Stabilization Fund, initially intended for

European reconstruction, to finance this covert operation, marking the beginning of the agency's involvement in electoral manipulation, a tactic that would be used repeatedly in subsequent years.

By the late 1940s and early 1950s, the CIA had cemented its role as the central intelligence agency of the United States, with a mission that extended far beyond intelligence gathering. The agency became a key player in the Cold War, tasked with conducting covert operations to counter Soviet influence around the globe. As the Cold War intensified, the CIA's role in shaping U.S. foreign policy became ever more significant, making it one of the most powerful and controversial institutions in the American government.

In 2004, the Intelligence Reform and Terrorism Prevention Act significantly altered the structure of the U.S. intelligence apparatus by creating the Office of the Director of National Intelligence (DNI). This new office took over many of the high-level responsibilities previously managed by the Central Intelligence Agency (CIA). The DNI was tasked with overseeing the entire United States Intelligence Community (IC), which includes 16 agencies, and managing the intelligence cycle, from information collection to analysis and dissemination. Among the critical functions transferred to the DNI was the responsibility for producing estimates that consolidated the views of all intelligence agencies and preparing intelligence briefings for the president.

On July 30, 2008, President George W. Bush further strengthened the DNI's role by issuing Executive Order 13470, which amended the longstanding Executive Order 12333. This move was intended to ensure that the DNI had the authority needed to fulfill its expanded oversight responsibilities.

Before establishing the DNI, the Director of Central Intelligence (DCI) had served as the head of the CIA and the president's principal intelligence advisor, overseeing the entire intelligence community. With the reforms introduced in 2004, the DCI's title changed to Director of the Central Intelligence Agency (D/CIA), and the CIA's reporting structure shifted. The CIA now reports directly to the DNI rather than to the president, although it still provides critical briefings to the president and congressional committees through the National Security Advisor, who remains a key member of the National Security Council. This council coordinates information from all intelligence agencies, including the National Security Agency (NSA) and the Drug Enforcement Administration (DEA).

In March 2015, the CIA underwent a major reorganization and reform initiative to position the agency for the future. In a public statement titled "Our Agency's Blueprint for the Future," the CIA announced several sweeping changes under the leadership of its director. One of the key reforms was the creation of a new Directorate of Digital Innovation tasked with developing cutting-edge digital technologies to keep the agency ahead of its adversaries. This new directorate also took on the responsibility of training CIA personnel in the use of these technologies and addressing emerging cyber threats, including cyberterrorism. As the primary cyber-espionage arm of the agency, the Directorate of Digital Innovation became central to the CIA's mission in the digital age.

Other significant changes included establishing a Talent Development Center of Excellence to enhance recruitment and training through the expansion of CIA University. A new office of the Chancellor was created to lead this university, unifying the agency's efforts in talent development. Additionally, the office of the Executive Director was empowered with a streamlined secretarial staff, allowing for more efficient day-to-day management of the CIA's operations.

Another key reform involved the creation of Mission Centers, each dedicated to a specific geographic region of the world. These centers were designed to combine the expertise of all five CIA directorates—Digital Innovation, Analysis, Operations, Science and Technology, and Support—under a unified command structure led by assistant directors. The goal was to foster collaboration and ensure that mission-specific objectives were met through the combined capabilities of the agency's various divisions.

In line with these changes, the CIA also reverted to some of its original organizational nomenclature. The National Clandestine Service, responsible for covert operations, was renamed the Directorate of Operations, while the Directorate of Intelligence was renamed the Directorate of Analysis, reflecting its core mission of providing detailed and insightful intelligence assessments to U.S. policymakers.

In the area of drone warfare, the CIA's role also saw fluctuations. Under President Barack Obama, the authority to conduct drone strikes was shifted from the CIA to the Department of Defense, placing these operations under military command. However, this policy was reversed by President Donald Trump, who restored the CIA's ability to carry out drone strikes targeting suspected terrorists, once again making the agency a key player in U.S. counterterrorism operations.

These changes reflect the CIA's ongoing evolution as it adapts to new technological, geopolitical, and operational challenges, while maintaining its central role in safeguarding U.S. national security.

The Defense Intelligence Agency

The Defense Intelligence Agency (DIA) plays a pivotal role in supporting the U.S. Space Force by providing military-focused intelligence, analyzing foreign space capabilities, and offering insights critical to national defense strategies in space. As a key member of the U.S. defense intelligence community, the DIA's mission is to provide military intelligence to warfighters, defense policymakers, and the broader intelligence community, and this role extends into the domain of space through its collaboration with the Space Force.

One of the DIA's core responsibilities is gathering and analyzing intelligence on foreign adversaries' space capabilities, including satellites, anti-satellite (ASAT) weapons, missile defense systems, and space-based assets that could threaten U.S. interests. By monitoring these developments, the DIA enables the Space Force to anticipate potential threats and vulnerabilities in U.S. space infrastructure, including reconnaissance, communications, and GPS satellites.

The DIA's intelligence assessments cover a wide range of topics relevant to the Space Force's mission, including the space doctrine and military strategies of adversaries like China and Russia, who are developing advanced space-based weaponry and counterspace technologies. These assessments help the Space Force shape its defensive and offensive strategies to maintain U.S. superiority in space, ensuring that America can protect its satellites, space assets, and broader interests in the event of a conflict.

Additionally, the DIA plays a key role in space situational awareness (SSA). By tracking and cataloging the movements of satellites and space debris, the DIA aids the Space Force in monitoring the operational environment of space, providing early warning of potential collisions, hostile actions, or disruptive activities by adversarial nations or other actors.

The DIA also works closely with the Space Force on counterintelligence and security efforts. Given the strategic importance of space technology, adversaries often seek to acquire sensitive U.S. space-related information through espionage. The DIA helps identify, track, and mitigate such efforts, providing vital intelligence on foreign attempts to infiltrate or undermine U.S. space capabilities.

Moreover, the DIA ensures that its intelligence contributions are integrated into joint military operations, including space-based defense initiatives. It collaborates with other intelligence agencies and military branches to provide a comprehensive intelligence picture, allowing the Space Force to make informed decisions on defending U.S. assets and maintaining dominance in space operations.

The Defense Intelligence Agency (DIA) is one of the primary intelligence and combat support agencies of the United States Department of Defense, specializing in military and defense intelligence. Established in 1961 by President John F. Kennedy under the direction of Secretary of Defense Robert McNamara, DIA plays a crucial role in informing both civilian and defense policymakers about the military intentions and capabilities of foreign governments and non-state actors. Its broad scope includes gathering and analyzing intelligence on foreign political, economic, industrial, and geographic factors, as well as health and medical information pertinent to national security.

Operating as a national-level organization, the DIA is structurally separate from other military intelligence components, although it integrates and coordinates intelligence efforts across the Department of Defense. The agency provides approximately one-quarter of the content for the President's Daily Brief, a critical document that keeps the President and top officials informed of global threats and developments. DIA's intelligence reach extends far beyond combat zones, with about half of its personnel stationed overseas in more than 140 countries, often serving within U.S. embassies. Its expertise in human intelligence (HUMINT), both overt and clandestine, and its management of military-diplomatic relations make DIA a pivotal player in the global intelligence community.

One of the agency's unique roles is serving as the national manager for Measurement and Signature Intelligence (MASINT), a highly technical form of intelligence that detects and identifies characteristics of various targets, including radar, acoustic, and nuclear signatures. It also oversees counterintelligence programs within the Department of Defense. Despite its wide-ranging intelligence functions, the DIA has no law enforcement authority, often a point of confusion in popular culture depictions.

Headquartered in Washington, D.C., at Joint Base Anacostia–Bolling, DIA's presence extends to the Pentagon and numerous Unified Combatant Commands. It coordinates with other government partners, including the CIA, deploying Defense Attaché Offices worldwide to represent U.S. defense interests in diplomatic relations and collect intelligence. Additionally, DIA personnel are based at several specialized centers and installations across the United States, such as the National Center for Medical Intelligence at Fort Detrick, the Missile and Space Intelligence Center in Huntsville, Alabama, and the Defense Intelligence Support Center in Reston, Virginia. The agency also administers the Intelligence Community Campus-Bethesda, which houses the National Intelligence University.

DIA's personnel consist of a diverse workforce of around 17,000, three-quarters of whom are career civilians with expertise in various military and defense intelligence fields. While no military background is required, nearly half of DIA employees have prior military experience. The agency marks the unclassified deaths of its personnel on a Memorial Wall, a tradition that reflects the inherent risks faced by its operatives.

DIA has played a significant role in U.S. intelligence operations throughout its history. During the Cold War, it provided vital intelligence to national security efforts, and its importance grew exponentially after the September 11 attacks, when the agency expanded both in size and operational scope. However, its involvement in controversial intelligence-gathering activities, including torture and domestic operations, has led to public scrutiny at times. The Director of the DIA, nominated by the President and confirmed by the Senate, serves as the primary intelligence advisor to the Secretary of Defense and the Director of National Intelligence. Additionally, the Director commands the Joint Functional Component Command for Intelligence, Surveillance, and Reconnaissance, part of the United States Strategic Command.

While less well-known than its civilian counterpart, the Central Intelligence Agency (CIA), DIA's focus on defense-related intelligence makes it a distinct entity within the U.S. Intelligence Community. Its responsibilities primarily cater to military needs at the national level, whereas the CIA serves broader intelligence objectives. DIA is also designated as a combat support agency, providing intelligence support directly to the Secretary of Defense, the Joint Chiefs of Staff, and Combatant Commanders. Though the media often portrays the relationship between the DIA and the CIA as competitive, the agencies share a collaborative partnership with distinct roles that complement one another. For example, CIA officers generally focus on broader intelligence matters, while DIA analysts are more concerned with military-specific intelligence, such as tracking missile movements or evaluating battlefield conditions.

The agency is divided into several directorates, each with a specific function. The Directorate for Operations oversees the Defense Clandestine Service (DCS), responsible for worldwide espionage activities, and the Defense

Attaché System, which coordinates military-diplomatic relations. The Directorate for Analysis provides all-source intelligence analysis, contributing to national security assessments such as the President's Daily Brief and National Intelligence Estimates. The Directorate for Science and Technology manages the agency's technical assets, particularly in MASINT and the Joint Worldwide Intelligence Communications System (JWICS), the primary secure network for classified intelligence. Lastly, the Directorate for Mission Services handles the agency's training centers and offers administrative and technical support for global operations.

DIA also manages several regional and functional centers focusing on different geographic areas and threats. These include the Americas and Transnational Threats Center, the Indo-Pacific Regional Center, and the Europe/ Eurasia Regional Center. Additionally, DIA oversees community-wide efforts, such as the National Center for Medical Intelligence and the Missile and Space Intelligence Center, which analyze threats from biological, chemical, and missile-related sources.

In 2021, DIA transitioned control of the National Intelligence University to the Office of the Director of National Intelligence. This educational institution, housed at the Intelligence Community Campus in Bethesda, Maryland, provides advanced training for intelligence professionals and maintains branch campuses at key military and intelligence locations.

From the tumultuous days of World War II until the dawn of the 1960s, the intelligence landscape within the United States military was fragmented and inefficient. Each of the three Military Departments—Army, Navy, and Air Force—operated their own intelligence apparatus, collecting, producing, and disseminating information for individual use. This siloed approach led to duplication of efforts, inflated costs, and, most critically, ineffective intelligence outputs. Often, conflicting estimates and assessments were presented to the Secretary of Defense and other federal agencies, undermining the reliability and coherence of national defense strategies.

Recognizing these significant shortcomings, the Defense Reorganization Act of 1958 sought to streamline military intelligence operations. However, the Act fell short of its objectives, leaving intelligence responsibilities muddled and coordination efforts inadequate. The result was a continued lack of national reliability and focus in intelligence matters. In response to these persistent issues, President Dwight D. Eisenhower appointed the Joint Study Group in 1960, tasking it with devising more effective organizational structures for the nation's military intelligence activities.

Acting upon the recommendations the Joint Study Group put forth, Defense Secretary Robert S. McNamara took decisive action in February 1961 by establishing the Defense Intelligence Agency (DIA). He directed the Joint Chiefs of Staff (JCS) to develop a comprehensive plan to integrate all military intelligence functions within the Department of Defense (DoD). This bold move was met with significant resistance from service intelligence units, whose commanders viewed the creation of DIA as an encroachment on their established domains. Despite the pushback, Air Force Lieutenant General Joseph Carroll emerged as a key figure in spearheading the new agency's planning and organization during the Cold War's heightened tensions, particularly surrounding the construction of the Berlin Wall.

On August 1, 1961, the Joint Chiefs of Staff issued Directive 5105.21, officially naming the new entity the Defense Intelligence Agency. By October 1 of the same year, DIA commenced operations with a modest staff operating out of borrowed office space. Initially reporting to the Secretary of Defense through the JCS, DIA was entrusted with the continuous collection, processing, evaluation, analysis, integration, production, and dissemination of military intelligence to support the DoD and other national stakeholders. The agency aimed to allocate intelligence resources more efficiently, manage DoD intelligence activities more effectively, and eliminate redundancies across facilities, organizations, and tasks.

Significant challenges and rapid developments marked the early years of DIA. In October 1962, just a year after its inception, DIA confronted its first major intelligence test during the Cuban Missile Crisis. The crisis unfolded when Soviet missiles were discovered in Cuba by Air Force reconnaissance planes, a discovery made possible by the analytical work of DIA analysts who had selected the flight paths that led to this critical intelligence breakthrough. This event underscored the agency's vital role in national security and validated its integrative approach to military intelligence.

Following the Cuban Missile Crisis, DIA continued to expand and refine its operations. In late 1962, the agency established the Defense Intelligence School, now known as the National Intelligence University, to train intelligence professionals. On January 1, 1963, DIA activated a new Production Center by merging several Service elements, consolidating operations at Arlington Hall Station, Virginia. The subsequent months saw the addition of an Automated Data Processing (ADP) Center in February, a Dissemination Center in March, and a Scientific and Technical Intelligence Directorate in April 1963. By July 1, 1963, DIA had assumed the staff support functions of the J-2, Joint Staff, further integrating its role within the DoD. Two years later, on July 1, 1965, DIA took over responsibility for the Defense Attaché System, transferring all intelligence functions from the individual Services to the new agency.

Throughout the 1960s, DIA analysts were deeply engaged in monitoring and evaluating a range of global developments. They focused on China's detonation of an atomic bomb and the subsequent Cultural Revolution, increasing unrest in African and South Asian nations, conflicts in Cyprus and Kashmir, and the pervasive missile gap between the United States and the Soviet Union. The decade also saw DIA navigating crises such as the Tet Offensive in Vietnam, the Six-Day War between Egypt and Israel, ongoing troubles in Nigeria, North Korea's seizure of the USS Pueblo, and the Warsaw Pact invasion of Czechoslovakia. These events tested DIA's ability to provide timely and accurate intelligence in a rapidly changing geopolitical landscape.

The early 1970s ushered in a transition period for DIA as the agency shifted its focus from merely consolidating its functions to establishing itself as a credible and reliable producer of national-level intelligence. This transition was complicated by significant manpower reductions between 1968 and 1975, which saw DIA's workforce decrease by 31 percent. These cuts necessitated mission reductions and a broad organizational restructuring, presenting formidable challenges as DIA sought to maintain its operational effectiveness. During this time, the agency had to contend with the rise of Ostpolitik in Germany, the emergence of the Palestine Liberation Organization in the Middle East, and U.S. military incursions into Cambodia from South Vietnam.

Despite these hurdles, DIA's reputation grew considerably by the mid-1970s. Decision-makers increasingly recognized the value of DIA's intelligence products, particularly as the agency's analysts concentrated on critical issues such as Lebanon, President Richard Nixon's historic visit to China, the 1973 Chilean coup d'état, the formation of Sri Lanka, and the plight of prisoners of war in Southeast Asia. The decade's latter part presented challenges, including détente, the development of arms control agreements, the Paris peace talks in Vietnam, the Yom Kippur War, and global energy concerns. However, intense congressional scrutiny during 1975–76, exemplified by the Murphy and Rockefeller Commission investigations into intelligence abuses, created turbulence within the Intelligence Community. These investigations ultimately led to Executive Orders that restructured Intelligence Community functions, further defining DIA's national and departmental responsibilities.

In the late 1970s, DIA adapted to these changes by reorganizing around five major directorates: production, operations, resources, external affairs, and J-2 support. This reorganization was a direct response to evolving intelligence needs and the necessity to enhance coordination and efficiency within the agency.

Entering the 1980s, the DIA had fully transformed into an integrated national-level intelligence agency. The release of its flagship publication, "Soviet Military Power," in 1981 marked a significant milestone. This comprehensive overview of Soviet military strength and capabilities was met with wide acclaim and became a serialized publication for the next decade, highlighting DIA's role in informing national defense strategies during the height of the Cold War. In 1983, under the Reagan Administration, DIA launched Project Socrates to research the flow of technology to the Soviet Union. Although Project Socrates was eventually terminated in 1990 due to funding cuts under the Bush Administration, its establishment underscored DIA's commitment to monitoring and countering technological advancements by adversarial nations.

The creation of the Clandestine Services organization, designated STAR WATCHER, in 1984 further solidified DIA's espionage capabilities. This unit collected intelligence on perceived conflict areas and potential adversaries in developing countries. STAR WATCHER aimed to establish a Joint Services career path for case officers, addressing inconsistencies in support for clandestine operations across the individual Services. This initiative was part of a broader effort to balance the CIA's focus on Soviet intelligence operations with a more diversified approach targeting third-world regions of potential military conflict.

Throughout the 1980s, DIA expanded its physical presence and operational capabilities. The agency moved into the newly constructed Defense Intelligence Agency Headquarters, a sprawling complex that became a central hub for DIA's activities. Military intelligence officer Gregory Davis bolstered efforts to establish a DoD-level espionage organization, who defined and established a clandestine services program under the U.S. Southern Command's "Plan Green." This program received authorization from JCS Chairman John Vessey and endorsement from the Senate Select Committee on Intelligence (SSCI), with significant sponsorship from Senators Jesse Helms and Barry Goldwater. The ensuing Goldwater–Nichols DoD Reorganization Act mandated that military officers serve in Joint Services assignments to qualify for flag rank, ensuring the continuity and development of DIA's case officers across the Services.

Designated as a combat support agency under the Goldwater–Nichols Act, DIA intensified its cooperation with Unified and Specified Commands and began developing a cohesive body of joint intelligence doctrine. Intelligence support to U.S. allies in the Middle East became increasingly critical as the Iran–Iraq War spilled into the Persian Gulf. DIA provided substantial intelligence support to Operation Earnest Will, closely monitoring incidents such as the Iraqi rocket attack on the USS Stark, the destruction of Iranian oil platforms, and Iranian assaults on Kuwaiti oil tankers. Additionally, DIA's analytical prowess was tested by conflicts like the "Toyota War" between Libya and Chad, turmoil in Haiti, and unrest in various parts of Latin America, Somalia, Ethiopia, Burma, Pakistan, and the Philippines.

By the close of the 1980s, DIA had firmly established itself as a cornerstone of the United States' national defense intelligence apparatus. Its comprehensive intelligence products, rigorous training programs, and strategic collaborations with other intelligence agencies positioned DIA as a pivotal player in navigating the complex and evolving challenges of the Cold War era.

Chapter 4 - The Russian Space Forces

The relationship between the United States Space Force and the Russian Space Forces has its origins rooted in decades of geopolitical competition that dates back to the Cold War. Emerging from the space race of the mid-20th century, the rivalry between the United States and the Soviet Union spurred technological advancements that would shape the modern landscape of military space operations. Today, that competition continues under different guises, with both nations possessing substantial capabilities in space and viewing the domain as critical to their national security interests.

The Cold War established the foundational dynamics of this relationship, with both superpowers striving for dominance in space as a means of demonstrating technological and military superiority. The Soviet Union's launch of Sputnik in 1957 shocked the world and ignited a sense of urgency within the United States, leading to the establishment of NASA and the eventual creation of military space programs within the Department of Defense. Over the following decades, space became an essential theater for surveillance, communications, and early warning systems, integral to both nations' strategic deterrence.

After the dissolution of the Soviet Union in 1991, Russia inherited the space assets and capabilities of its predecessor. Despite economic struggles and reduced funding, Russia maintained a prominent position in military space operations, continuing to develop reconnaissance satellites, missile detection systems, and communications networks. The Russian Space Forces, officially formed in 1992, were a direct response to the changing geopolitical landscape, tasked with managing the nation's military and strategic interests in space.

Meanwhile, the United States continued to invest heavily in space technologies throughout the 1990s and early 2000s, recognizing the increasing reliance on space-based assets for national defense and global military operations. The 21st century brought new challenges as space became more congested, contested, and competitive, with emerging powers like China adding complexity to the traditional US-Russia dynamic. It was in this evolving context that the United States created the Space Force in December 2019, elevating space to an independent military domain.

The relationship between the US Space Force and the Russian Space Forces today is characterized by both competition and a recognition of mutual interests in maintaining space as a functional and secure domain. Both nations possess highly capable military space assets, including satellite constellations for communications, navigation, missile warning, and intelligence gathering. However, their approaches to space operations reflect the broader tensions in their geopolitical relationship.

Russia, for instance, has been increasingly focused on developing counter-space capabilities aimed at undermining the United States' technological advantage. Anti-satellite (ASAT) weapons, cyber operations targeting space infrastructure, and the development of jamming and spoofing technologies are all part of Russia's strategy to challenge US dominance in space. In 2021, Russia's ASAT test that destroyed one of its defunct satellites created a cloud of space debris, prompting international condemnation, particularly from the United States. Such actions highlight the growing concern over space becoming a battlefield, with the risk of escalation extending beyond the Earth's atmosphere.

Conversely, the US Space Force has focused on enhancing the resilience of its space assets, increasing the security of its satellite networks, and ensuring that the United States maintains a technological edge in space-based capabilities. Collaboration with the intelligence community, private sector, and international allies plays a significant role in the US Space Force's strategy to secure and defend space.

Despite these competitive dynamics, the US and Russia maintain limited cooperation in space, particularly in the realm of civilian space exploration. The International Space Station (ISS), a symbol of post-Cold War cooperation, continues to operate as a joint endeavor, with astronauts from both nations working side by side. However, as geopolitical tensions escalate on Earth, there is growing concern that such collaboration may not extend into future space activities.

In conclusion, the relationship between the US Space Force and the Russian Space Forces is a complex mixture of historical rivalry and present-day competition, shaped by decades of space exploration and military advancement. Both nations see space as critical to their national defense strategies, but the increasing militarization of space raises the possibility of confrontation. As the geopolitical landscape evolves, the actions of both the US and Russia will likely determine whether space remains a realm of cooperation or becomes a battlefield in the ongoing struggle for global power.

The Russian Space Forces (Космические войска России, Kosmicheskie vodka Rossii, KV) are the space branch of the Russian Aerospace Forces. They were reestablished on August 1, 2015, following the merger of the Russian Air Force and the Russian Aerospace Defence Forces. This merger restored the space forces as a distinct entity after their dissolution in 2011. Historically, the Russian Space Forces have played a pivotal role in both military and space exploration efforts, reflecting Russia's continued emphasis on the strategic importance of space.

The Russian Space Forces were initially formed on August 10, 1992, shortly after the establishment of the Russian Armed Forces. This made them the first independent space force in the world, a unique distinction at the time. Initially, the Space Forces shared control of the Baikonur Cosmodrome with Roscosmos, the Russian Federal Space Agency, and also operated the Plesetsk and Svobodny Cosmodromes. However, their independence was short-lived. In July 1997, the Russian Space Forces were dissolved and incorporated into the Strategic Missile Forces, which aligned space operations more closely with missile and defense programs.

In response to organizational needs, the Russian Space Forces were reformed as an independent branch on June 1, 2001, as part of a military reorganization. However, this iteration of the Space Forces lasted only a decade, and in December 2011, they were once again dissolved and replaced by the Russian Aerospace Defence Forces, a move intended to consolidate aerospace and defense capabilities under a single command. Four years later, on August 1, 2015, the Russian Aerospace Forces were formed through the merger of the Air Force and Aerospace Defence Forces. As a result, the Russian Space Forces were reestablished as one of the three sub-branches within the new military structure.

The roots of Russian space forces can be traced back to the Soviet era, where space exploration was fundamentally linked to military objectives. The Soviet Strategic Missile Forces had integrated space operations as early as February 12, 1955, when the Central Committee of the Communist Party of the Soviet Union and the Council of Ministers issued a joint executive order to establish a Scientific Research and Testing Range. This site, known by the codename Tayga Installation (Обект «Тайга»), was officially designated as the 5th Scientific Research and Testing Range (5-й Научно-исследовательский испытательный полигон, 5-й НИИП) on June 2, 1955. Located in the Kyzylorda Region of the Kazakh Soviet Socialist Republic, the range was central to developing the Soviet space program and its strategic ballistic missile force.

Unlike the United States, where NASA, a civilian agency, led space exploration, the Soviet Union placed space exploration under military control. The Soviet space program was closely tied to the development of intercontinental ballistic missiles (ICBMs), under the direction of the Strategic Missile Forces (RVSN). The first commander of these space troops, Marshal of Artillery Mitrofan Ivanovich Nedelin, later became Chief Marshal of Artillery and was instrumental in advancing both missile and space technologies.

The Baikonur Cosmodrome, established as the central hub for Soviet space activities, played a key role in the launch of the first human into space. On April 12, 1961, Yuri Gagarin's historic flight aboard Vostok 1 marked a significant milestone in the Soviet space program. To maintain secrecy, Soviet authorities referred to the launch site as Baikonur, after a nearby railway station, although its official designation was Tyuratam.

The military dimension of the Soviet space program was evident in the establishment of various operational units within the 5th Scientific Research and Testing Range. These units were tasked not only with space exploration but also with the development and testing of ICBMs. For example, the 43rd Separate Engineer Testing Detachment, originally part of the 627th Missile Regiment, played a key role in testing the R-16U ICBM, achieving the first launch from an underground shaft in 1962. Similarly, the 676th Missile Regiment was transferred to the range to conduct tests on the R-9A ICBM, demonstrating the close integration of space and missile programs.

Throughout the 1960s, the Soviet space forces tested and launched various missiles and spacecraft. They also conducted high-profile demonstration launches for foreign dignitaries, such as the 1966 launch witnessed by French President Charles de Gaulle. By the end of the decade, the Soviet space forces had established themselves as a key component of both military strategy and space exploration.

In 1964, the Soviet Ministry of Defense established the Central Directorate for Space Assets of the Missile Troops of Strategic Purpose (Центральное управление космических средств Ракетных войск стратегического назначения, ЦУКОС РВСН). This Directorate was responsible for coordinating the various space-related units within the military, consolidating the command structure for space operations. By 1970, it had been upgraded to the Main Directorate for Space Assets (Главное управление космических средств, ГУКОС РВСН), reflecting its growing importance.

The Plesetsk Cosmodrome, established in 1957, became a key site for testing the R-7 ICBM, one of the Soviet Union's most important missile systems. This site, along with the Command and Measurement Complex for Control of Space Vehicles, established in the same year, ensured that the Soviet space forces had the infrastructure necessary to support both military and scientific space missions.

Following the dissolution of the Soviet Union, the newly formed Russian Federation sought to continue the legacy of its predecessor's space achievements. The reformation of the Russian Space Forces in 1992 marked a new chapter in Russian space and military operations. Despite organizational changes over the years, the Russian Space Forces remain a crucial component of Russia's national defense and space exploration strategy.

Today, as one of the sub-branches of the Russian Aerospace Forces, the Russian Space Forces are responsible for securing Russia's interests in space. Their duties include managing satellite systems, monitoring missile threats, and maintaining space superiority. The reestablishment of the Russian Space Forces in 2015 underscores the continued importance of space in modern military operations, positioning Russia as a key player in the global space race.

In 1981, the Soviet Union reorganized its space defense assets by removing the Main Directorate from the Strategic Missile Forces (RVSN) and placing it under the direct control of the General Staff. This change marked the transformation of the Directorate from the Strategic Missile Forces (ГУКОС РВСН) to the General Staff (ГУКОС ГШ ВС СССР), solidifying its elevated role in military space operations. Five years later, in 1986, the Main Directorate for Space Assets underwent further restructuring, becoming the Directorate of the Chief of Space Assets under the Ministry of Defence, reflecting its increasing importance in Soviet defense strategy.

The collapse of the Soviet Union in 1992 brought further changes. The Directorate of the Chief of Space Assets was reformed into a separate branch known as the Military Space Forces (Военно-космические силы, ВКС). This marked a pivotal moment when space defense officially became a distinct branch of the armed forces. It is crucial to distinguish between the Military Space Forces (ВКС) established at this time and the later formation of the Aerospace Forces, which also shared the ВКС acronym but represented a broader integration of air and space defense capabilities.

The Soviet Union's efforts to protect against missile attacks and monitor space activities began in earnest on March 30, 1967, with the creation of the Directorate of the Chief of Anti-Missile and Anti-Space Defense. This unit, commanded by Lieutenant-General Yuri Votintsev, fell under the Soviet Air Defence Forces. By 1971, the 1st Division for Warning Against Missile Attack (1-я Дивизия предупреждения о ракетном нападении, WAMA) had been established, with its headquarters in Solnechnogorsk. This division operated radar stations in Olenegorsk, Murmansk Oblast, and Skrunda, Latvian SSR.

The 2nd Division for Space Observation (2-я Дивизия разведки космического пространства) followed in 1973, with its headquarters at the Serpukhov-15 site. It managed space observation facilities across the USSR, including the 145th Center for Space Control in Noginsk-9, near Moscow, and radar stations in Irkutsk and Kazakhstan. These divisions began a sophisticated space control network to monitor space objects and potential missile threats.

In 1965, the Soviet Union initiated the formation of a comprehensive Space Control System (Система контроля космического пространства, ЦККП). The 45th Division for Space Control, established in 1966 under the Soviet Air Defence Force, was crucial in monitoring space activities from its base in the closed military town of Noginsk-9. Over the following decade, these units expanded, forming the backbone of Soviet space defense capabilities.

By 1977, the 1st Division WAMA was reorganized into the 3rd Separate Army of Special Purpose for Warning Against Missile Attack, which absorbed the space observation units. The creation of the 9th Separate Corps for Anti-Missile Defense in 1978, based in Akulovo, Moscow Oblast, further bolstered the USSR's anti-missile capabilities.

In the late 1980s, the Soviet space and missile defense landscape continued to evolve. In 1988, the 45th Division was upgraded to the 18th Separate Corps for Space Control, though it was later reduced back to a division in the 1990s. Despite this downsizing, the unit remained operational, eventually becoming the Main Centre for Reconnaissance of Situation in Space, a key player in monitoring space activities.

The renaming of the Anti-Missile and Anti-Space Defense Troops to the Missile and Space Defense Troops in 1992 signified another shift in focus, reflecting the growing importance of space defense in post-Soviet Russia. In 1998, the 9th Separate Corps for Anti-Missile Defense was restructured and transferred from the Air Defence Forces to the Missile Troops of Strategic Purpose, further cementing the integration of missile and space defense under a unified command.

In 2001, the Missile and Space Defense Troops and the Military Space Troops were detached from the Strategic Missile Forces and formed into a separate entity—the Space Forces (Космические войска). This move aimed to refocus Russia's efforts on space-based capabilities. Colonel General Anatoly Perminov was appointed the first commander of the newly formed Space Forces, followed by Generals Vladimir Popovkin and Oleg Ostapenko. This independent branch remained active until 2011, playing a crucial role in space defense and operations.

Finally, in 2015, the Space Forces were absorbed into the newly established Russian Aerospace Forces (Воздушно-космические силы), an entity designed to integrate air and space defense into a single structure. This transition marked the culmination of decades of evolution, as the Soviet and later Russian military adapted to the increasing importance of space in national defense.

The Russian Space Forces, established as a distinct branch of the Russian military, play a vital role in defending the nation against missile attacks, maintaining control over military space assets, and supporting the broader aerospace capabilities of Russia. Their primary responsibilities include providing early warnings of missile launches, managing ballistic missile defense, and deploying and maintaining space vehicles in orbit, such as the GLONASS global positioning system and reconnaissance satellites like Persona.

One of the core missions of the Space Forces is to alert political leaders and military commanders of impending missile attacks. The Division of Warning of Missile Attack, headquartered in Solnechnogorsk, serves as the hub for this crucial task. It operates a sophisticated network of radar stations, such as the Dnepr radar systems—also known as "Hen House"—which are located in Olenegorsk (Russia), Beregovo and Sevastopol (Ukraine), Mishelevka

(Irkutsk), and Balkhash (Kazakhstan). Additional radar sites, like the Daryal system in Pechora (Russia) and Gabala (Azerbaijan), enhance the early warning capability by providing extensive coverage of potential missile launches.

Complementing the early warning systems, the Russian Space Forces manage the A-135 anti-ballistic missile system, which is designed to protect Moscow from missile attacks. This advanced system includes radar installations and missile sites, such as those at Novopetrovska, Klin, and Korolev, equipped with interceptors like the 51T6 and 53T6 missiles. These assets form part of the 9th Division of Defense Against Missiles, headquartered in Sofrino, which operates radar stations such as the Don-2N system (also known as "Pill Box") and Dunay-3U in Chekhov-7.

In addition to missile defense, the Space Forces are responsible for deploying and controlling in-orbit space vehicles. This includes the GLONASS global navigation satellite system, which became fully operational in October 2010. As a key component of Russia's military infrastructure, GLONASS provides precise geolocation services worldwide. By 2008, under the leadership of Colonel General Vladimir Popovkin, the Space Forces had placed 18 GLONASS satellites in orbit, achieving full operational status two years later.

The Space Forces also operate the Titov Main Test and Space Systems Control Centre, responsible for monitoring and maintaining Russia's military satellites and space systems network. This includes tracking and managing reconnaissance satellites, communication satellites, and other strategic space assets essential for national defense.

The Space Forces' ability to monitor activities in space is another critical function. The Optical-Electronic Complex "Okno" (Window), located near Nurek, Tajikistan, was commissioned in 2002. This facility can track space objects up to 40,000 kilometers from Earth. The Okno system monitors satellites, debris, and other space objects using telescope-like equipment housed in large spheres, much like the U.S. GEODSS system. This complex is essential in space surveillance, giving the Russian military real-time data on space activities.

In addition to the Okno complex, the Space Forces operate the Krona radar system in Zelenchukskaya, which uses both optical and laser systems to track objects in low Earth orbit. Radar installations at Sofrino, Balkhash, and Mishelevka further enhance the military's ability to maintain space situational awareness.

At the heart of the Space Forces' command structure is the 3rd Missile-Space Defence Army, which integrates various divisions responsible for missile defense and space control. The 1st Division of Warning of Missile Attack and the 9th Division of Defense Against Missiles are among the most important units, both operating out of Solnechnogorsk and Sofrino, respectively.

The 45th Division of Space Control, based in the Noginsk area, oversees Russia's space surveillance efforts, including the operation of the Okno complex and other radar sites across the country. This division is crucial for maintaining control over Russia's space assets and ensuring that any potential threats from space are detected and tracked.

In 2018, the Russian Space Forces were integrated into the newly formed 15th Aerospace Forces Army. This reorganization aimed to strengthen the coordination between air defense and space operations under a unified command structure. Key installations under the 15th Army include the Titov Main Test and Space Systems Control Centre, the Main Centre for Missile Attack Warning, and the Main Space Intelligence Centre. These facilities form the backbone of Russia's modern aerospace defense network.

The Plesetsk Cosmodrome, one of Russia's primary launch facilities for military and civilian satellites, also falls under the Space Forces' jurisdiction. From Plesetsk, the Space Forces launch satellites into orbit and test new missile technologies, ensuring the continuous expansion and modernization of Russia's space capabilities.

The officer ranks of the Russian Space Forces were structured similarly to those of other branches of the Russian military, with a clear hierarchy that defined responsibilities and command. These ranks were categorized into three main groups: general officers, senior officers, and junior officers.

General and Flag Officers:

Генера́л а́рмии (General of the Army): The highest rank among general officers, reserved for senior military leaders with significant command responsibilities.

Генера́л-полко́вник (Colonel General): A senior rank in the general officer category, often commanding large military units or divisions.

Генера́л-лейтена́нт (Lieutenant General): Responsible for overseeing large formations or providing high-level strategic oversight.

Генера́л-майо́р (Major General): The entry-level rank for general officers, typically commanding smaller formations or acting as deputies in larger units.

Senior Officers:

Полко́вник (Colonel): A senior field officer, often commanding a regiment or brigade, with significant strategic and tactical responsibilities.

Подполко́вник (Lieutenant Colonel): Typically serves as a deputy commander of a regiment or brigade or holds a staff position within a higher command structure.

Майо́р (Major): Responsible for commanding a battalion or serving in senior staff roles within a regiment.

Junior Officers:

Капита́н (Captain): Commands a company or serves in staff positions; this rank is the first level of company leadership.

Ста́рший лейтена́нт (Senior Lieutenant): A junior officer rank that typically involves leading smaller units or acting as second-in-command within a company.

Лейтена́нт (Lieutenant): The entry-level officer rank, responsible for leading a platoon or serving in other junior command roles.

Мла́дший лейтена́нт (Junior Lieutenant): Often newly commissioned officers tasked with commanding smaller units or taking on staff duties.

Курса́нт (Cadet): Designates officers-in-training who have yet to receive a commission.

Other Ranks

The non-commissioned officers (NCOs) and enlisted personnel ranks of the Russian Space Forces followed a similar structure to other branches, with distinctions made between senior NCOs and junior enlisted personnel. These ranks provided the backbone for operational and logistical support in the Space Forces.

Senior NCOs:

Ста́рший прапо́рщик (Senior Praporshchik): The highest NCO rank, typically responsible for overseeing the training and discipline of enlisted personnel.

Прапо́рщик (Praporshchik): A warrant officer rank, often serving as a technical expert or senior advisor in specific operational areas.

NCOs and Enlisted:

Старшина́ (Starshina): Equivalent to a master sergeant, responsible for unit discipline and the training of lower-ranked soldiers.

Ста́рший сержа́нт (Senior Sergeant): Acts as a platoon sergeant or similar, leading squads or sections.

Сержа́нт (Sergeant): A junior NCO, typically responsible for leading a squad or section.

Мла́дший сержа́нт (Junior Sergeant): The entry-level NCO rank, often tasked with smaller leadership roles within a unit.

Ефре́йтор (Private First Class): An experienced enlisted rank, acting as a team leader or assistant to an NCO.

Рядово́й (Private): The most junior enlisted rank, typically responsible for basic operational duties within a unit.

The leadership of the Russian Space Forces has been defined by several key figures, each contributing to the development and evolution of the branch. Here are notable commanders throughout its history:

Kerim Kerimov (Lieutenant General): As one of the pioneers of Soviet space defense, Kerimov played a crucial role in the early years of the Space Forces, serving from 1964 to 1965.

Andrei Karas (Colonel General): Commanded from 1965 to 1979, overseeing significant advancements in missile defense and space operations during the Cold War.

Aleksandr Maksimov (Colonel General): His tenure from 1979 to 1989 coincided with the modernization of Soviet space assets.

Vladimir Ivanov (Colonel General): Led the Space Forces from 1989 to 1996 during a period of transition following the collapse of the Soviet Union.

Anatoly Perminov (Colonel General): Took command in 2001 and was instrumental in reorganizing the Space Forces as an independent branch.

Vladimir Popovkin (Colonel General): Commanded from 2004 to 2008, focusing on modernizing the Russian space infrastructure.

Oleg Ostapenko (Colonel General): Served from 2008 until the dissolution of the Space Forces in 2011, overseeing their integration into the broader Aerospace Forces.

Aleksandr Golovko (Colonel General): Since August 2015, Golovko has led the newly established Russian Aerospace Forces, which includes the Space Forces.

The Russian Space Forces operated a sophisticated network of installations and radars designed to protect the nation from missile attacks and to monitor space activities. These capabilities were critical for early missile warning, space surveillance, and missile defense.

Key Installations and Assets:

Pechora Radar Station: This radar facility was crucial for monitoring missile launches and providing early warning to Russian forces.

A-135 Anti-Ballistic Missile System: Located near Moscow, this system was designed to intercept and destroy incoming ballistic missiles aimed at the Russian capital.

Peresvet Anti-Air Laser Combat System: A cutting-edge system developed to protect strategic missiles from potential threats.

Okno (Window) Optical Tracking Facility: Situated near Nurek, Tajikistan, this facility tracked space objects up to 40,000 kilometers away, providing crucial space situational awareness.

GLONASS Global Positioning System: A satellite navigation system managed by the Space Forces, it became fully operational in 2010 and serves both military and civilian purposes.

By 2018, the Space Forces had become part of the larger 15th Aerospace Forces Army, which included the Titov Main Test and Space Systems Control Centre, the Main Centre for Missile Attack Warning, and the Main Space Intelligence Centre. The Plesetsk Cosmodrome, one of Russia's primary launch sites, also fell under the jurisdiction of the Space Forces.

Chapter 5 - The People's Liberation Army

The geopolitical relationship between the United States Space Force and China's People's Liberation Army (PLA), particularly the PLA Strategic Support Force (PLASSF), is shaped by growing strategic competition, mistrust, and the recognition that space is a key domain for future military dominance. Unlike the United States' long-standing rivalry with Russia in space, the US-China dynamic is relatively newer but no less consequential. In recent years, China has emerged as a major player in space exploration and military space capabilities, positioning itself as a direct competitor to the United States in the quest for technological supremacy and control over the high frontier.

The PLA's interest in space as a domain of warfare began to take shape in the 1990s when China recognized the importance of space-based assets in modern warfare, particularly after observing the United States' reliance on satellite systems during the Gulf War. The Chinese military leadership realized that space assets provided critical advantages in communications, navigation, intelligence gathering, and precision targeting. In response, China began investing heavily in its own space program, leading to the establishment of the PLASSF in 2015. This branch of the PLA is responsible for operations in space, cyber, and electronic warfare, reflecting China's understanding that future conflicts will be multi-domain and highly reliant on space-based technologies.

The United States views China's rapid advancements in space with growing concern, especially in light of China's stated goal of becoming a global leader in space by 2049, coinciding with the centenary of the founding of the People's Republic of China. China's space ambitions include not only civilian achievements, such as the Chang'e lunar exploration program and the Tianhe space station, but also significant military advancements. The PLASSF oversees the deployment of China's satellite constellations, including those for missile early warning, intelligence gathering, communications, and navigation, all of which have military applications.

One of the most alarming aspects of China's space strategy from the US perspective is its development of counter-space capabilities. China has made significant progress in anti-satellite (ASAT) weapons, which have the potential to disrupt or destroy US satellites that are crucial for military and civilian operations. In 2007, China tested an ASAT missile that destroyed one of its weather satellites, creating a massive debris field and sending a clear signal of its counter-space capabilities. Since then, China has continued to develop technologies that could challenge the United States' dominance in space, including direct ascent ASAT missiles, co-orbital satellites capable of disabling or interfering with other satellites, and ground-based jamming systems targeting satellite communications and GPS signals.

The United States, particularly through the newly established Space Force, has responded by emphasizing the need to protect its space assets and maintain superiority in the domain. The Space Force's mission includes not only the defense of US satellites but also the ability to project power in space, ensuring freedom of operation in a domain increasingly seen as vital to national security. As part of this strategy, the US has focused on strengthening the resilience of its space infrastructure, investing in advanced missile detection systems, and collaborating with private space companies to enhance satellite capabilities.

The rivalry between the US Space Force and the PLASSF is not only about military space assets but also about the broader struggle for technological leadership. Space has become an arena for technological competition between the two nations, with China advancing rapidly in areas such as satellite quantum communications, reusable launch systems, and space exploration. The successful launch of China's space station, Tiangong, and its long-term plans for lunar exploration are part of a broader strategy to challenge the United States' leadership in space.

Unlike the US-Russia relationship in space, which is built on decades of cooperation and competition, the US-China relationship is marked by deeper mistrust. The United States views China's military space ambitions as part of a broader effort to challenge American global dominance. At the same time, China sees US space capabilities as a potential threat to its rise as a global superpower. This mutual suspicion has made direct cooperation in space limited, and while both nations participate in international frameworks such as the Outer Space Treaty, their strategic postures suggest a growing competition in the militarization of space.

The geopolitical relationship between the United States Space Force and China's People's Liberation Army is defined by increasing competition for control and dominance over space. Both nations re

The People's Liberation Army (PLA) serves as the military arm of the Chinese Communist Party (CCP) and the People's Republic of China. It encompasses four main service branches—Ground Force, Navy, Air Force, and Rocket Force—as well as four distinct operational arms: the Aerospace Force, Cyberspace Force, Information Support Force, and Joint Logistics Support Force. The PLA operates under the leadership of the Central Military Commission (CMC), with the chairman acting as commander-in-chief, reflecting its unique relationship with the ruling party.

The People's Liberation Army (PLA) is the military force of the People's Republic of China and plays a critical role in the country's political and defense structure. It is housed in the Ministry of National Defense compound, known as the "August 1st Building," symbolizing its central position in the nation's military organization. The PLA is a key component of China's broader armed forces, including the People's Armed Police (PAP), reserves, and militia forces.

Under the Chinese Communist Party (CCP), the doctrine of "the Party must always control the gun" (Chinese: ◇◇◇◇; pinyin: Dǎng zhǐhuī qiāng) reinforces the political control of the military. This doctrine ensures that the CCP maintains absolute authority over the armed forces, a foundational principle of the Chinese military structure. The PLA and PAP have the largest delegation in the National People's Congress (NPC), with servicemembers elected by military election committees within top-level military subdivisions, such as theater commands and service branches. At the 14th National People's Congress, the PLA's joint delegation included 281 deputies, representing over 9% of the total, and all of whom were CCP members.

The Central Military Commission (CMC) is the highest body governing the PLA. Under the arrangement of "one institution with two names," both a state CMC and a Party CMC exist, although both have identical personnel, organization, and functions, effectively operating as a single entity. The only time the membership between the two bodies differs is during the transition period between the Party National Congress and the subsequent National People's Congress, when personnel changes occur.

The CMC comprises a chairman, vice chairpersons, and regular members. The chairman, who also serves as the commander-in-chief of the PLA, is typically China's paramount leader. Since 1989, this position has been held alongside the role of CCP general secretary. Notably, while the Ministry of National Defense acts as a diplomatic arm of the CMC, it holds no command authority over the military, thereby insulating the PLA from external influence. However, the Minister of Defense is always a member of the CMC.

Prior to 2016, the PLA was governed by four general departments: the General Political, General Logistics, General Armament, and General Staff Departments. However, as part of Xi Jinping's military reforms, these departments were abolished and replaced with 15 functional departments that report directly to the CMC. Among them, the Discipline Inspection Commission plays a crucial role in rooting out corruption within the military.

The CCP maintains absolute control over the PLA, ensuring that military personnel are thoroughly indoctrinated in the Party's ideology through regular political education. The political commissar system operates throughout the military, with CCP committees and political commissars present at regiment levels and higher. These political commissars, alongside military commanders, share equal authority, ensuring that CCP ideology is upheld in all military decisions. Battalion- and company-level units also maintain political directors and instructors, responsible for implementing Party decisions and maintaining discipline.

CCP committees make key decisions at each level of the military, reinforcing the Party's absolute leadership over the PLA. Non-CCP organizations are strictly forbidden within the military, and only the CCP can appoint military leaders at all levels.

The PLA operates under a grade system that determines command hierarchies from the CMC down to the platoon level. Entities command those of lower grades and coordinate with those of equal grades. Since 1988, all PLA organizations, positions, and officers have been assigned a grade that dictates authority, pay, eligibility for certain roles, and retirement age.

This grading system extends beyond the military, as it parallels the civilian grade system, facilitating coordination between military and civilian entities. Career progression within the PLA often includes lateral transfers between positions of the same grade, though such moves are not considered promotions.

Historically, the grade or position of an officer was more important than their military rank, with multiple ranks coexisting within a single grade. Ranks primarily served as visual indicators of an officer's position, especially when interacting with foreign personnel. However, reforms initiated in 2021 aimed to shift the military towards a more rank-centric system. By 2023, a revised grade structure associated one rank per grade, though some ranks continued to span multiple grades, reflecting the ongoing transformation of the PLA's command system.

In the PLA's etiquette, addressing personnel by position rather than rank was preferred, reflecting the importance of organizational hierarchy over personal titles. These reforms underscore China's ongoing efforts to modernize and professionalize its military, aligning its structure more closely with international standards while maintaining the CCP's absolute control over the armed forces.

The PLA's operational control is a complex structure divided between the service headquarters and theater commands, with responsibilities shifting depending on whether the country is at peace or engaged in war. This structure is designed to ensure regional and joint-service coordination, reinforcing China's ability to respond to various military and geopolitical challenges.

The five theater commands of the PLA represent joint, multi-service organizations tasked with overseeing military strategy, planning, tactics, and policy specific to their designated geographic areas. Each theater command coordinates forces from the army, navy, air force, and other service branches to manage defense and combat operations. During wartime, these commands assume full control of subordinate units, whereas in peacetime, they share control with the respective service headquarters. Force-building and modernization, however, remain the responsibility of the service branches and the Central Military Commission (CMC).

The five theater commands, in order of stated significance, are as follows:

Eastern Theater Command: Responsible for key areas like Taiwan and the East China Sea, this command is considered the most strategically significant due to its proximity to contested territories and regions of heightened military interest.

Southern Theater Command: Covers China's southern borders, including the South China Sea, where China asserts expansive territorial claims.

Western Theater Command: The largest geographically, covering Tibet, Xinjiang, and China's borders with India and other western neighbors, making it critical for border security.

Northern Theater Command: Focuses on security issues near North Korea and Russia, managing threats from the Korean Peninsula and maintaining defense relations with northeastern neighbors.

Central Theater Command: Centrally located, this command is responsible for the security of the capital, Beijing, and the surrounding regions, playing a pivotal role in homeland defense.

These theater commands were established in 2016 as part of sweeping military reforms aimed at improving joint operational capability and shifting from the army-centric structure of previous military regions. Unlike military regions, which were primarily administrative in peacetime and relied on the army for wartime coordination, the theater commands are designed to function as fully integrated, multi-service operational commands during both peace and war.

In addition to the theater commands, each service branch of the PLA retains certain operational control responsibilities. For example:

The PLA Army headquarters is responsible for critical regions such as the Beijing Garrison, Tibet Military District, and Xinjiang Military District, as well as China's border and coastal defenses.

The PLA Navy controls China's counter-piracy patrols in the Gulf of Aden, reflecting the navy's broader role in maritime security and international operations.

The Joint Staff Department (JSD) nominally oversees operations beyond China's immediate periphery, though this predominantly applies to army-led operations.

The overlapping responsibilities of the theater commands and service branches often result in the need for arbitration by the CMC to resolve jurisdictional disputes, as both operate at the same grade level within China's military hierarchy.

The Central Military Commission (CMC) maintains ultimate authority over all military operations in China, including those of the theater commands and service branches. The CMC oversees 15 departments, commissions, and offices that directly manage and support the PLA's activities, including:

Joint Staff Department: Responsible for the overall strategic planning and operational coordination of the PLA.

Political Work Department: Oversees the CCP's political influence within the military.

Logistic Support Department: Manages the supply chains and resources essential for military readiness.

Equipment Development Department: Oversees the acquisition and development of advanced military hardware.

Training and Administration Department: Responsible for ensuring that PLA personnel are trained to the highest standards.

National Defense Mobilization Department: Coordinates military readiness in response to national emergencies.

In addition, the Discipline Inspection Commission plays a key role in maintaining internal discipline within the PLA, rooting out corruption and ensuring adherence to CCP principles.

The PLA consists of several service branches, each critical to China's overall defense strategy:

PLA Ground Force: The largest and oldest branch of the Chinese military, focusing on land-based military operations.

PLA Navy: Responsible for safeguarding China's maritime interests and securing its coastline and overseas trade routes.

PLA Air Force: Provides air defense and power projection capabilities across China and beyond.

PLA Rocket Force: Manages China's strategic missile arsenal, including its nuclear deterrent.

Various research institutes, such as the Academy of Military Science, the National Defense University, and the National University of Defense Technology, support these branches and drive technological innovation and military research. These institutions contribute to developing new strategies, weapons systems, and defense technologies.

The reforms initiated by Xi Jinping in 2015 were intended to streamline the PLA's structure, reduce redundancy, and improve joint-service operations. These reforms included the replacement of military regions with theater commands and the dissolution of the army-dominated General Staff Department. The goal was to create a modernized, combat-ready force capable of addressing contemporary security challenges. The reforms also emphasized shifting to a rank-centric system, enhancing the importance of military rank in promotions and decision-making processes.

Several state-owned enterprises (SOEs) in China have established internal People's Armed Forces Departments overseen by the People's Liberation Army (PLA). These internal units collaborate with grassroots organizations to gather intelligence, identify potential security risks, and neutralize threats in their early stages. According to the People's Liberation Army Daily, this proactive role in intelligence gathering and security management aims to prevent issues from escalating into larger security concerns.

The PLA's education system includes two top-tier academic institutions directly subordinate to the Central Military Commission (CMC): the National Defense University and the National University of Defense Technology. These institutions are critical for military education and leadership training in China. Additionally, the PLA oversees 35 military-affiliated academic institutions tied to its various service branches and arms, while the People's Armed Police (PAP) operates seven affiliated institutions. These academic bodies are vital in producing well-trained, highly skilled military personnel and fostering research and innovation in military science and technology.

The PLA comprises four main services: the Ground Force, Navy, Air Force, and Rocket Force. Collectively, these services form the backbone of China's military power. The PLA has undergone significant modernization and restructuring in recent decades to better align with contemporary military strategies and geopolitical goals.

The PLA Ground Force (PLAGF) is the largest branch of the PLA, with 975,000 active-duty personnel, representing roughly half of the PLA's total manpower. The PLAGF is organized into 12 group armies, numbered from the 71st to the 83rd, distributed among the PLA's five theater commands. In wartime, reserve forces, including infantry and anti-aircraft artillery divisions, can be mobilized to support these active units. The Ground Force has been streamlined in recent years, with a significant portion of non-combat personnel reduced, allowing more resources to modernize air, naval, and missile forces. Commander Liu Zhenli and Political Commissar Qin Shutong lead the Ground Force.

The PLA Navy (PLAN) has evolved from a subordinate branch of the PLA Ground Force into a major force in its own right. With 300,000 personnel, the PLAN is organized into three fleets: the North Sea Fleet (headquartered in Qingdao), the East Sea Fleet (headquartered in Ningbo), and the South Sea Fleet (headquartered in Zhanjiang). These fleets include surface ships, submarines, naval aviation, coastal defense, and marine units.

The PLAN also includes a Marine Corps of 25,000 personnel and a Naval Aviation Force of 26,000 personnel, equipped with attack helicopters and fixed-wing aircraft. Since the 1990s, the navy has undergone rapid modernization and is now transitioning to a blue-water navy, capable of projecting power globally. In 2012, then-CCP General Secretary Hu Jintao expressed the ambition to build China into a "strong maritime power." As of 2024, the PLAN is numerically the largest navy in the world, according to the U.S. Department of Defense. The PLAN is led by Commander Dong Jun and Political Commissar Yuan Huazhi.

The PLA Air Force (PLAAF), with 395,000 personnel, is organized into five Theater Command Air Forces (TCAF) and was originally structured into 24 air divisions. However, as of 2024, the system has been restructured, converting most divisions into air brigades, with the larger bomber divisions and special mission units remaining intact. Each air brigade or division is equipped with 24 to 50 aircraft, making the PLAAF one of the most capable air forces in the world. The PLAAF also operates a robust surface-to-air missile (SAM) Corps and three airborne divisions.

China's stealth fighter programs, referred to as J-XX and XXJ by Western intelligence agencies, reflect ongoing advancements in developing fifth-generation fighter jets. Commander Chang Dingqiu and Political Commissar Guo Puxiao lead the PLAAF.

The PLA Rocket Force (PLARF) is China's strategic missile force, responsible for managing both nuclear and conventional missile arsenals. With at least 120,000 personnel, the Rocket Force operates six operational missile bases (numbered from 61 through 66), each assigned to specific theater commands. Base 67 serves as China's central nuclear weapons storage facility. China's nuclear arsenal is estimated to consist of between 100 and 400 thermonuclear warheads. Commander Li Yuchao and Political Commissar Xu Zhongbo lead the PLARF.

In addition to the four main services, the PLA operates four specialized arms: the Aerospace Force, Cyberspace Force, Information Support Force, and Joint Logistics Support Force. These arms were established in April 2024 to address modern warfare's increasingly technological and network-centric nature. They reflect China's focus on integrating advanced technologies and cyber capabilities into its military strategy, ensuring that the PLA remains capable of defending against emerging threats in space, cyberspace, and information warfare.

The PLA's modernization and restructuring efforts demonstrate China's commitment to transitioning from a land-based, ground force-centric military to a modernized, technologically advanced force capable of projecting global power. The strategic shift toward naval, air, and missile forces underscores the country's growing focus on asserting its influence in key areas such as the South China Sea and beyond, while also maintaining a strong emphasis on technological superiority in aerospace and cyberspace domains.

Initially, the PLA was an all-volunteer force, but the Military Service Law of 1955 introduced compulsory military service as part of China's effort to modernize its armed forces. Since the late 1970s, the PLA has evolved into a hybrid force, combining both conscripts and volunteers. Under current regulations, conscripts who complete their service obligation may continue their military careers as volunteers, serving up to 16 years in total. While military service is de jure obligatory for all Chinese citizens, mandatory conscription has not been enforced since the founding of the People's Republic of China in 1949. Instead, the PLA relies on a volunteer-based system, with recruitment efforts focused on attracting both men and women into its ranks.

Women have played a significant role in the PLA throughout its history, particularly during revolutionary periods such as the Chinese Civil War (1927–1949) and the Second Sino-Japanese War (1937–1945). During these conflicts, women often participated in unconventional warfare, including combat roles. However, after the establishment of the People's Republic of China, the PLA's transition toward a conventional military organization reduced women's roles primarily to support positions, including medical, logistics, and administrative functions.

Despite this shift, military service remained a prestigious choice for women, offering education, training, and upward mobility opportunities. During the Cultural Revolution, military service became a path to avoid political repression, further solidifying its importance in Chinese society. In the 1980s, the PLA underwent significant demobilization as part of broader economic reforms, reducing female participation. However, by the 1990s, the PLA began recruiting women again, primarily in non-combat roles, such as communications, intelligence, and cultural work.

Women in the PLA often hold specialized positions and are more likely to serve as cadets or officers than enlisted personnel due to their academic and professional qualifications. Although traditionally, women were excluded from combat roles, this began to change in the 2010s, with women increasingly serving in mixed-gender units and being held to the same physical standards as their male counterparts. Women are celebrated for their contributions to the military, and achievements are highlighted through state media, including their participation in peacekeeping missions, service on aircraft carriers, and performance in special forces units. Despite the PLA not releasing detailed statistics on gender composition, it is estimated that approximately 5% of the active military force is female.

Promoting national unity and territorial integrity has been a central theme in the history of the PLA, particularly in the recruitment of ethnic minorities. The PLA has actively recruited minority groups since its early years, forming specialized units such as Mongol cavalry brigades during the Chinese Civil War and welcoming ethnic Korean volunteers during the Korean War.

In the 1950s, ethnic minorities were given preferential treatment, with efforts made to recruit and train officers from minority backgrounds. For example, in the Inner Mongolia military region, ethnic Mongols made up 52%

of all officers. The PLA also emphasized "socialist culture" and assimilation policies during periods such as the Great Leap Forward and the Cultural Revolution, fostering a sense of shared identity among soldiers from diverse backgrounds.

Ethnic minority officers are often assigned to positions in their home regions, and many have risen to prominent ranks in the PLA. For instance, over 34% of battalion and regimental cadres in the Yi autonomous region militia were of Yi ethnicity, while 45% of militia cadres in Tibetan regions were of Tibetan ethnicity. However, representation at the highest levels of the PLA remains limited for ethnic minorities, with most holding junior officer positions.

Prominent minority figures in PLA history include Ulanhu, an ethnic Mongol who served in high-ranking roles in Inner Mongolia and as Vice President of China, and Saifuddin Azizi, an ethnic Uyghur Lieutenant General who held a position in the CCP Central Committee. While there have been a few cases of ethnic tensions within the PLA, such as the defection of ethnic Tatar General Margub Iskhakov to the Soviet Union in the 1960s, such incidents have been rare and often linked to broader political dissatisfaction rather than ethnic divisions.

In contemporary times, ethnic minorities remain an integral part of the PLA, with the majority serving in junior officer roles. However, few minorities ascended to the highest ranks, whereas Han Chinese still dominated leadership positions. The PLA continues to promote ethnic diversity within its ranks, aligning recruitment efforts with state policies emphasizing national unity and the integration of ethnic groups into Chinese society.

The rank structure of the People's Liberation Army (PLA) reflects the organizational complexity and hierarchical nature of China's military forces. Each branch of the PLA—Ground Force, Navy, Air Force, Rocket Force, and Strategic Support Force—has its own set of ranks for officers, non-commissioned officers (NCOs), and enlisted personnel. This system, heavily influenced by Soviet military tradition, establishes clear distinctions between different levels of command and responsibility, ensuring an organized chain of command.

The PLA's officer ranks are categorized into three main groups: General, Senior, and Junior. The structure is consistent across the branches, though with slight variations in terminology to reflect the specific roles of the Ground Force, Navy, Air Force, Rocket Force, and Strategic Support Force.

At the highest level of command, Generals (◇◇, Shàngjiàng) hold the ultimate leadership positions across all branches. These officers are responsible for overseeing large military divisions and strategic operations. Below them are Lieutenant Generals (◇◇, Zhōngjiàng) and Major Generals (◇◇, Shàojiàng), who command corps and divisions, respectively.

While the rank titles are mostly uniform across the branches, the PLA Rocket Force and Strategic Support Force have specialized ranks reflecting their unique operational focus. These branches manage China's strategic missile forces and information warfare capabilities, respectively. For example, the Rocket Force follows a similar rank structure as the Ground Force but focuses on commanding strategic missile operations.

The PLA Strategic Support Force also follows this hierarchy, managing cyber, space, and electronic warfare. This relatively new branch highlights China's increasing focus on modern warfare capabilities that extend beyond traditional military operations.

The People's Liberation Army's rank structure is a comprehensive system maintaining order and discipline across its diverse branches. Each role is clearly defined from the highest-ranking generals to the newest enlisted personnel, ensuring that the PLA operates as an efficient and cohesive military force. By maintaining consistency across the branches, with only minor adjustments to reflect branch-specific duties, the PLA's hierarchical system is built to sustain the organization in the face of both traditional and modern military challenges.

As China has steadily increased its military funding, the development of advanced weapons technologies has become a focal point of its defense strategy. According to the United States Department of Defense, China invests heavily in cutting-edge technologies such as kinetic-energy, high-powered lasers, microwave, particle-beam, and electromagnetic pulse (EMP). These advancements reflect China's ambition to modernize its military capabilities and establish itself as a formidable global power in defense technology.

However, China's rapid modernization has drawn scrutiny from international observers. The PLA has countered claims that its military advancement is largely dependent on the acquisition of advanced technologies from foreign nations. Senior Chinese officials have stated that such accusations are politically motivated and undermine the country's reputation. Nevertheless, reports suggest that China has indeed benefited from technology transfers, acquiring European diesel engines for its warships, military helicopter designs from Eurocopter, and French anti-submarine sonar and helicopter technology. Other contributions include Australian technology for the Houbei-class missile boat and Israeli-supplied American missile and aircraft technology.

China's role as a global arms supplier has expanded significantly over the past two decades. According to data from the Stockholm International Peace Research Institute (SIPRI), China became the world's third-largest exporter of major arms between 2010 and 2014, marking a 143% increase from the previous period (2005-2009). By 2020, China had surpassed Russia to become the second-largest arms exporter in the world. China's share of global arms exports grew from 3% to 5%, supplying major arms to 35 countries.

China's key export partners during 2010-2014 were Pakistan, Bangladesh, and Myanmar, which together received over 68% of Chinese arms exports. Additionally, China made significant arms deals with countries such as Venezuela, Algeria, Indonesia, and Nigeria, supplying a variety of military hardware ranging from armored vehicles and aircraft to anti-ship missiles and unmanned aerial vehicles (UAVs).

While China's arms industry has advanced rapidly, allowing it to reduce its dependency on foreign imports, it continued to rely on Russia and Ukraine for critical components, particularly engines for combat aircraft and naval vessels. Between 2010 and 2014, China's arms imports from Russia, France, and Ukraine decreased by 42%, signaling its growing self-sufficiency. Nevertheless, China still imported large numbers of engines for indigenous aircraft and ships, with many of these engines produced under long-standing agreements with British, French, and German manufacturers.

In August 2021, China demonstrated its significant progress in hypersonic weapons technology by testing a nuclear-capable hypersonic missile that circled the globe before descending toward its target. This test, which reportedly caught U.S. officials by surprise, underscored China's advances in hypersonic technology, positioning it as a leader in this emerging field of military capability.

China's military advancements were also showcased during joint exercises with Russia, such as the 2021 Exercise Zapad-81. Chinese forces participated with novel military equipment, including the KJ-500 airborne early warning and control aircraft, J-20 and J-16 fighters, Y-20 transport planes, and an array of surveillance and combat drones. Another notable joint exercise occurred in August 2023 near Alaska, further demonstrating China's military coordination with Russia and its global reach.

China's military strategy increasingly includes cyberwarfare, an asymmetric tool that allows the PLA to engage adversaries without direct armed conflict. Western military analysts believe that China has been conducting cyber espionage and cyber-attacks on other countries since 1999, with these operations increasing over time.

Cyberwarfare has proven effective in modern conflict as it targets critical infrastructure, information systems, and strategic assets. Two Chinese colonels, Qiao Liang and Wang Xiangsui, outlined this approach in their book Unrestricted Warfare, which argues that non-military methods, such as cyber-attacks, can achieve strategic military goals without the need for traditional armed confrontation.

In 2011, the PLA formally acknowledged its cyber capabilities, which were further highlighted in 2013 when a hacker group known as "Comment Crew" was linked to China's military. In May 2014, a U.S. Federal Grand Jury indicted five PLA officers on charges of cyber espionage against U.S. companies, and in 2020, several PLA members were charged with hacking into the Equifax database as part of a massive data theft operation. Despite China's denial of these claims, the incidents highlight its growing capabilities in cyberwarfare.

China's nuclear weapons program dates back to its first nuclear test in 1964, followed by its first hydrogen bomb test in 1967. Despite its participation in the Comprehensive Nuclear-Test-Ban Treaty (CTBT) in 1996, China has yet to ratify the treaty, and the exact size of its nuclear arsenal remains a closely guarded state secret. Various estimates place the number of nuclear warheads at approximately 438 as of 2024, with the U.S. Department of Defense suggesting the number could exceed 500, making China the third-largest nuclear power globally.

China's nuclear doctrine has historically been centered on a policy of no first use, with a focus on maintaining a secure second-strike capability. This strategy aims to deter potential adversaries through the threat of retaliation in the event of a nuclear attack.

China does not underestimate the importance of space to modern military operations. After observing the pivotal role space assets played in the U.S. military's success during the Gulf War, China has placed a strong emphasis on space as a key domain in conflict and global strategic competition. The PLA operates numerous satellite constellations designed for reconnaissance, navigation, communication, and counterspace operations.

China's space-based reconnaissance capabilities rely on Jianbing (vanguard) satellites, known publicly as Yaogan and Gaofen satellites, which collect electro-optical, radar, and electronic intelligence. These assets are vital to China's military strategy, providing targeting information and enabling secure communications for its forces.

The PLA has also developed anti-satellite and counterspace capabilities, conducting its first successful anti-satellite missile test in 2007. More recently, China has invested in hypersonic space vehicles, including the Shenglong Spaceplane, which represents a significant leap in China's ability to conduct space-based operations.

As China continues to assert its presence in space, the PLA plays an essential role in the nation's space exploration efforts. All Chinese astronauts to date have been selected from the PLA Air Force, and China became the third country to independently send humans into space with the 2003 flight of Yang Liwei aboard the Shenzhou 5 spacecraft. Since then, China has made rapid progress, with several successful crewed space missions, reflecting the country's growing ambitions in space exploration and warfare.

The PLA's origins can be traced back to the left-wing units of the National Revolutionary Army (NRA) during China's Republican Era. In 1927, these units broke away during an uprising against the nationalist government and formed the Chinese Red Army. During the Second Sino-Japanese War, the Chinese Red Army was reabsorbed into the NRA under two specific units—the New Fourth Army and the Eighth Route Army. However, in 1947, as civil war resumed in China, these communist military forces were reconstituted as the People's Liberation Army, setting the stage for their crucial role in the founding of the People's Republic of China in 1949.

Since its inception, the PLA has employed a series of evolving military strategies, referred to as "strategic guidelines." Of the nine strategies developed since 1949, the most significant changes came in 1956, 1980, and 1993, each reflecting China's shifting geopolitical and domestic priorities. The PLA's role in national politics is substantial, with a large presence in the National People's Congress (NPC) through its joint delegation, which includes 281 deputies, all of whom are members of the CCP. This highlights the PLA's deep integration into the Chinese political system, making it not merely a state military but a crucial pillar of the CCP's governance.

The PLA differs from traditional nation-state militaries, as it is directly linked to the CCP rather than the government of China. Its primary mission is not only national defense but also the defense of the party itself. As the guarantor of CCP control, the PLA's loyalty to the party is paramount. The CMC exercises supreme command over the armed forces, with the CCP general secretary also holding the position of CMC chairman—a combination of roles that consolidates the party's influence over the military. While the Ministry of National Defense functions as a liaison between the military and external entities, it holds no direct command authority over the PLA, ensuring that the party retains firm control over military operations.

Geographically, the PLA organizes its forces into five theater commands, each assigned to a specific region within China. This structure enables the efficient deployment of military resources in both peacetime and conflict. As of 2023, the PLA stands as the world's largest military force in terms of personnel and operates with the second-largest defense budget globally, amounting to US$296 billion. This represents 12 percent of worldwide defense expenditures, underscoring China's rising military prominence. The PLA is also one of the fastest-modernizing forces, with advancements in defense technology and power projection capabilities, positioning it as a potential global military superpower.

The PLA's influence extends beyond traditional military functions, particularly in its involvement in maritime territorial disputes. In these scenarios, the PLA Navy plays a pivotal role in coordinating with the China Coast Guard, a branch of the People's Armed Police (PAP), to protect China's claimed interests in contested waters. This integration of military and paramilitary forces illustrates the PLA's adaptability in both wartime and peacetime operations, reinforcing its critical role in China's strategic posture on the global stage.

The primary mission of the People's Liberation Army (PLA) is the defense of the Chinese Communist Party (CCP) and its core interests. The PLA is not merely a national military force but serves as the guarantor of the CCP's survival and rule. As such, the CCP prioritizes maintaining control over the PLA and ensuring its unwavering loyalty. In 2004, paramount leader Hu Jintao articulated the PLA's mission in a comprehensive statement that outlined its essential duties: to safeguard the leadership of the CCP, protect the sovereignty, territorial integrity, and internal security of the People's Republic of China, defend the country's national development, and preserve global peace. China characterizes its military strategy as one of "active defense," which was officially defined in a 2015 white paper as "We will not attack unless we are attacked, but we will surely counterattack if attacked."

The origins of the PLA can be traced to August 1, 1927, during the Nanchang Uprising, an event that marked the beginning of the Chinese Civil War. Following the Shanghai Massacre earlier that year, communist elements within the National Revolutionary Army (NRA) rebelled against the Kuomintang (KMT), under the leadership of key figures such as Zhu De, He Long, Ye Jianying, and Zhou Enlai. This military force, initially known as the Chinese Workers' and Peasants' Red Army—commonly referred to as the Red Army—was the predecessor of the modern PLA.

The Red Army faced numerous campaigns launched by Chiang Kai-Shek's KMT forces during the mid-1930s, including the devastating battles that culminated in the Long March of 1934–1935. This arduous retreat became a defining moment in CCP history, as the Red Army survived despite being pursued by superior KMT forces.

During the Second Sino-Japanese War (1937–1945), the Red Army was nominally integrated into the NRA, forming two distinct units: the Eighth Route Army and the New Fourth Army. Throughout this period, communist forces employed guerrilla tactics against Japanese invaders, while simultaneously expanding their influence by recruiting KMT troops and paramilitary units operating behind enemy lines. By doing so, the CCP consolidated its military strength even as it avoided direct large-scale engagements with Japanese forces.

Following Japan's surrender in 1945, the CCP maintained the NRA's unit structures until a significant military reorganization took place in February 1947. At that time, the Eighth Route Army and the New Fourth Army were merged to form a new force: the People's Liberation Army (PLA). By the end of 1948, this newly organized force, which had grown to over a million soldiers, was instrumental in winning the Chinese Civil War and establishing the People's Republic of China in 1949.

Following the establishment of the People's Republic of China, the PLA underwent significant reorganization to enhance its capabilities and structure. In November 1949, the leadership framework for the Air Force was created, and the Navy's command structure followed in April 1950. These foundational changes were critical in expanding the PLA's role beyond a traditional ground force.

In 1950, the PLA further diversified by establishing leadership for artillery, armored units, air defense forces, public security troops, and worker-soldier militias. Over time, specialized forces such as chemical warfare defense units, railroad forces, communications, and strategic forces—including engineering, logistics, and medical services—were also created, adding layers of capability to the growing military apparatus.

In its early years, the PLA was composed mainly of peasants, and its organizational culture was deeply egalitarian. This peasant composition represented a radical break from the hierarchical norms that dominated Chinese society at the time. Formal military ranks were not introduced until 1955, which contributed to an egalitarian spirit within the ranks, fostering a sense of solidarity between officers and soldiers. According to sociologist Alessandro Russo, this break from traditional hierarchies created unprecedented forms of social equality within the PLA.

In the formative years of the People's Republic of China, the PLA played a dominant role in shaping the nation's foreign policy. As a newly established military force emerging from years of civil war and conflict with foreign powers, the PLA's involvement in military and political strategy was pivotal. During this period, it was not only tasked with defending China's borders but also with extending the CCP's influence domestically and internationally. This close alignment between military objectives and political authority laid the foundation for the PLA's enduring role as both a military and political institution within China.

In the 1950s, the People's Liberation Army (PLA) embarked on a significant transformation, evolving from a largely peasant force into a modern military. With substantial assistance from the Soviet Union, the PLA began modernizing its capabilities, adopting advanced military strategies, equipment, and organizational structures. Since 1949, the PLA has utilized nine distinct military strategies, referred to as "strategic guidelines." The most critical shifts occurred in 1956, 1980, and 1993, reflecting China's evolving military and geopolitical priorities.

Part of this modernization involved the reorganization of the PLA into thirteen military regions in 1955, which helped streamline command structures and enhance operational efficiency. The PLA's modernization was quickly tested when, in November 1950, units operating under the name of the People's Volunteer Army intervened in the Korean War. As United Nations forces under General Douglas MacArthur advanced towards the Yalu River, Chinese forces launched a counteroffensive that drove them out of North Korea and captured Seoul. However, the Chinese forces were eventually pushed back north of the 38th Parallel. The Korean War served as a catalyst for the rapid modernization of the People's Liberation Army Air Force (PLAAF), solidifying China's resolve to build a capable, modern military force.

In 1962, the PLA was again engaged in a significant conflict, this time with India during the Sino-Indian War. In this border war, Chinese forces inflicted heavy losses on Indian troops and secured territorial gains. However, subsequent border clashes in 1967 resulted in significant tactical and numerical losses for the PLA. These conflicts underscored the need for continued modernization and reform within the PLA, particularly in areas related to strategy and battlefield tactics.

During the Cultural Revolution (1966–1976), the PLA underwent considerable upheaval. Military region commanders, who had previously held their positions for extended periods, saw their tenures disrupted. High-profile commanders such as Xu Shiyou, Yang Dezhi, Chen Xilian, and Han Xianchu, who had served in their respective military regions for over a decade, were replaced or removed. The early days of the Cultural Revolution also saw the abandonment of military ranks, which had been introduced in 1955 to professionalize the PLA, reflecting the broader anti-hierarchical sentiment of the time.

The drive to modernize the PLA gained new momentum with the announcement of the Four Modernizations by Premier Zhou Enlai, a program later championed by Deng Xiaoping. Among these, the modernization of China's military was prioritized to equip the PLA with modern weapons and advanced military doctrine. In line with Deng's reforms, the PLA demobilized millions of personnel starting in 1978, shifting its focus towards recruitment, manpower efficiency, and the implementation of modern education and training methods. This shift aimed to enhance the professionalism and combat readiness of the force.

In 1979, the PLA fought Vietnam in the Sino-Vietnamese War, a conflict sparked by border skirmishes. Both sides claimed victory; however, Western analysts largely concluded that Vietnam had outperformed the PLA, exposing weaknesses in China's military capabilities. Despite the war's outcome, the conflict spurred further reform and modernization efforts within the PLA.

During the Sino-Soviet split, China's relations with the Soviet Union became increasingly strained, leading to border clashes in the late 1960s. Both nations backed each other's adversaries in conflicts around the globe. As tensions escalated, China adopted a cautious stance towards Afghanistan, maintaining neutral relations during the reign of Afghanistan's monarch. However, following the 1978 communist coup in Afghanistan, relations quickly deteriorated. The pro-Soviet Afghan government supported China's adversaries in Vietnam and accused China of backing anti-communist militants in Afghanistan.

When the Soviet Union invaded Afghanistan in 1979, China responded by providing significant support to the Afghan mujahideen fighting against Soviet forces. This support included military training and vast quantities of arms, including anti-aircraft missiles, rocket launchers, and machine guns. China's PLA Ground Force played a critical role in training the mujahideen, even relocating training camps from Pakistan into China itself. Chinese military advisors and troops assisted the Afghan resistance, while China also bolstered its own military presence in Xinjiang, anticipating potential Soviet incursions into Chinese territory.

As the PLA transitioned into the late 20th century, it continued its modernization journey, integrating lessons learned from its various military conflicts. China's military posture of "active defense" persisted, ensuring that while China would not strike first, it would vigorously defend its territorial and strategic interests when challenged.

In 1981, the People's Liberation Army (PLA) conducted its largest military exercise since the founding of the People's Republic of China. Held in North China, this exercise marked the PLA's increasing focus on modernization and training to meet new challenges.

During the 1980s, the PLA gained more autonomy and was permitted to engage in commercial activities, allowing it to offset budgetary limitations in exchange for reducing its political role. This downsizing freed up resources for China's economic development, but also introduced systemic corruption due to a lack of oversight and ineffective self-regulation. Corruption within the ranks persisted into the late 2010s, significantly lowering the PLA's operational readiness and proficiency, and creating barriers to modernization. Jiang Zemin's attempts to divest the PLA from commercial ventures were only partially successful, as many of these businesses remained controlled by associates of high-ranking officers. These issues eroded party control over the military and contributed to inefficiencies. Beginning in the early 2010s, Xi Jinping's anti-corruption campaigns and military reforms were aimed at addressing these long-standing problems and reasserting the CCP's authority over the PLA.

Following the suppression of the 1989 Tiananmen Square protests, the PLA temporarily emphasized ideological correctness, reinforcing its political loyalty to the Chinese Communist Party (CCP). However, the PLA soon resumed its focus on reform and modernization, which remained a priority as China looked to develop a more capable and advanced military force.

Starting in the 1980s, the PLA began a gradual shift from a land-based force centered on ground troops to a more agile, mobile, and high-tech military capable of conducting operations beyond China's borders. This shift was driven by the understanding that a large-scale land invasion, such as from Russia, was no longer the primary threat. Instead, the PLA recognized the potential for conflicts over Taiwan's independence, especially if supported by the United States, or territorial disputes in the South China Sea, particularly regarding the Spratly Islands.

In 1985, under the leadership of the Central Military Commission (CMC) and the CCP, the PLA shifted its strategic focus. Moving away from the previous doctrine of preparing to "hit early, strike hard, and fight a nuclear war," the PLA began developing in an era of peace. Deng Xiaoping stressed that China's military needed to prioritize quality over quantity. By 1987, the PLA had reduced its numbers by one million, with leadership staffing cut by about 50 percent. Further reductions occurred during the Ninth Five-Year Plan (1996–2000), with another 500,000 personnel cut, followed by an additional 200,000 in the early 2000s. This downsizing allowed the PLA to focus on increasing mechanization and informatization, aiming to be capable of fighting high-intensity wars in the future.

The PLA also began integrating lessons from global conflicts. The 1991 Gulf War, in particular, revealed the PLA's technological shortcomings, prompting Jiang Zemin to order a decade-long modernization program. This program aimed to enhance the military's capabilities in response to concerns about potential conflict with Taiwan and increasing tensions in the region. Additionally, the PLA closely studied the successes and failures of U.S. military operations in conflicts such as the Kosovo War, the invasions of Afghanistan and Iraq, and the Iraqi insurgency. These observations reinforced China's need to shift from a numbers-dominated ground force to a more sophisticated, high-tech military.

In 1993, Jiang Zemin formally incorporated the concept of a "revolution in military affairs" (RMA) into China's national military strategy. The goal was to transform the PLA into a force capable of winning "local wars under high-tech conditions." This shift required a focus on short, decisive campaigns with limited geographic scope and political objectives, contrasting with the large-scale, protracted wars of the past. More attention was given to areas like reconnaissance, mobility, and deep reach. This new vision also prioritized resources for the navy and air force, while preparing the PLA for potential space warfare and cyber-warfare.

From 2002 onwards, the PLA began participating in military exercises with foreign armed forces, a trend that grew in frequency through the 2010s. Between 2018 and 2023, over half of these exercises focused on non-traditional military missions such as anti-piracy and anti-terrorism operations, typically involving combatting non-state actors. Notably, in 2009, the PLA conducted its first military exercise in Africa, a humanitarian and medical training exercise in Gabon.

Throughout the past two decades, the PLA has acquired advanced weapon systems from Russia, including Sovremenny-class destroyers, Sukhoi Su-27 and Su-30 fighter aircraft, and Kilo-class diesel-electric submarines. Simultaneously, China has developed its own advanced military technology, including the Type 052D class guided-missile destroyers and the Chengdu J-10 fighter jet. In further demonstrating its technological ambitions, China also developed the Chengdu J-20 stealth fighter and launched Jin-class nuclear submarines capable of delivering nuclear strikes across the Pacific.

By 2023, the PLA boasted a fleet of three aircraft carriers, the latest of which, the Fujian, was launched in 2022. China's expansion of naval and air capabilities reflects its ongoing efforts to project power regionally and globally.

In addition to its modernization efforts, the PLA has participated in various United Nations peacekeeping operations and humanitarian missions. The largest of these was the deployment of 524 medical personnel between 2014 and 2015 to combat the Ebola outbreak in West Africa. The PLA has also provided engineers, logistical units, and paramilitary forces for peacekeeping missions in countries such as Lebanon, the Republic of the Congo, Sudan, and more recently, Mali and South Sudan.

From 2015 to 2016, the PLA underwent a comprehensive reorganization to enhance joint operations and streamline command structures. The PLA Ground Force, Rocket Force, and Strategic Support Force were established as new units, while the CMC replaced the traditional military departments with newly formed bodies to oversee different operational aspects. The PLA also reorganized its command structure by replacing seven military regions with five theater commands: Northern, Southern, Western, Eastern, and Central. This reorganization aimed to integrate different military branches and improve overall combat readiness and operational coordination.

The PLA's commitment to modernization and reform was further demonstrated during its 90th-anniversary celebrations in 2017, when it held its largest military parade to date outside of Beijing at the Zhurihe Training Base in Inner Mongolia. This parade highlighted the PLA's achievements and its continued efforts to build a world-class military force.

In December 2023, a major leadership purge occurred within the PLA, with several high-ranking generals removed from their posts in the National People's Congress, reflecting ongoing efforts to address corruption and maintain the party's control over the military.

China's military presence has expanded globally, with overseas deployments including bases in Djibouti, Tajikistan, and a listening station in Cuba. The PLA has also operated in Cambodia and maintains the Espacio Lejano Station in Argentina, operated by a PLA unit. As a member of the United Nations, China has consistently contributed to peacekeeping operations, demonstrating its growing global influence and military capabilities.

The PLA's transformation into a modern, high-tech military has been a central focus of Chinese defense policy since the 1980s. With advanced weapons systems, restructured command structures, and increased global involvement, the PLA continues to evolve as a key instrument of China's geopolitical ambitions.

Since its founding in 1949, the People's Republic of China has been involved in various military engagements and conflicts, both within its borders and on the international stage. These conflicts reflect China's evolving geopolitical interests and military strategy throughout the 20th and 21st centuries.

China's modern military history begins with the Chinese Civil War (1927–1950), a brutal conflict between the nationalist Kuomintang (KMT) and the communist forces led by Mao Zedong. This war culminated in the establishment of the People's Republic of China in 1949, following the communist victory. During this period, China also fought the Second Sino-Japanese War (1937–1945), aligning with the Allies during World War II to resist Japanese occupation, which severely impacted the country's population and infrastructure.

In 1949, the Yangtze Incident occurred when British warships traveling on the Yangtze River were attacked by People's Liberation Army (PLA) forces, further showcasing the new government's assertiveness on Chinese territory. That same year, the People's Republic of China consolidated its control over the western region of Xinjiang, incorporating it into the country.

In the early 1950s, China focused on consolidating its territorial integrity. This effort included the Annexation of Tibet in 1950, a controversial move that has influenced Chinese relations with its neighbors and the international community ever since.

China's involvement in the Korean War (1950–1953) marked its first major military engagement on the international stage. Fighting under the banner of the Chinese People's Volunteer Army, Chinese forces intervened to support North Korea against United Nations and South Korean forces, signaling China's emergence as a significant player in Cold War geopolitics.

The 1950s also saw China engaged in two crises over Taiwan. The First Taiwan Strait Crisis (1954–1955) and the Second Taiwan Strait Crisis (1958), involving military confrontations around the islands of Quemoy and Matsu, showcased ongoing tensions between China and the nationalist government in Taiwan.

China's engagement in the Vietnam War (1955–1970) further solidified its involvement in Cold War conflicts. Although China's role was more indirect, offering substantial material and logistical support to North Vietnam, it was a critical component of China's strategy to counter Western influence in Asia.

In 1962, China fought the Sino-Indian War, a brief but intense conflict over disputed territories in the Himalayan region. This war resulted in a decisive Chinese victory but left a lasting legacy of tension between the two nations. Further border skirmishes with India occurred in 1967, continuing the disputes over the boundaries between the two countries.

Tensions with the Soviet Union also escalated in the late 1960s, leading to the Sino-Soviet border conflict of 1969. This brief yet severe military clash nearly brought the two communist giants to full-scale war, highlighting ideological and territorial disagreements between China and the Soviet Union.

In 1974, China engaged in the Battle of the Paracel Islands with South Vietnam, a confrontation over control of these strategic islands in the South China Sea. The victory cemented China's hold over the islands, a source of ongoing disputes in the region.

The Sino-Vietnamese War of 1979 marked China's last major direct military conflict. The brief but bloody war was launched in response to Vietnam's invasion of Cambodia, with China seeking to "teach Vietnam a lesson." Although the conflict ended without clear gains for either side, it underscored China's willingness to use force to influence regional politics.

Since 1979, China has largely avoided large-scale wars but has been involved in several Sino-Vietnamese conflicts (1979–1990), including skirmishes along the border that reflected ongoing tensions with its southern neighbor. In 1988, the Johnson South Reef Skirmish with Vietnam further demonstrated China's determination to assert its territorial claims in the South China Sea.

Domestically, the PLA played a critical role in enforcing martial law during the Tiananmen Square protests of 1989, a pivotal moment in modern Chinese history, marked by the violent suppression of pro-democracy demonstrators in Beijing.

In the 1990s, China reasserted its sovereignty over Hong Kong and Macau. The PLA established the Hong Kong Garrison in 1997 following the handover of the territory from British control, and similarly established a garrison in Macau in 1999 after its return from Portuguese administration.

International Peacekeeping and Modern Conflicts (2000–Present)

In the 21st century, China's military engagements have shifted focus from direct warfare to peacekeeping and international collaboration. Since 2007, Chinese forces have participated in UNIFIL peacekeeping operations in Lebanon, and since 2009, they have taken part in anti-piracy operations in the Gulf of Aden, reflecting China's growing role in international security.

China also contributed to global efforts by participating in the search and rescue operation for Malaysia Airlines Flight 370 in 2014 and continues to engage in UN peacekeeping missions in Mali (2014) and South Sudan (2015).

Recently, renewed tensions with India have led to the China–India skirmishes of 2020–2021, as the two nations once again clashed over their disputed Himalayan border.

As of early 2024, China has not engaged in a major war since 1979, instead focusing on relatively minor military engagements and expanding its influence through international peacekeeping and strategic regional conflicts. These developments reflect China's broader shift toward asserting itself as a global power while maintaining stability within its borders.

Chapter 6 - The National Aeronautics and Space Administration (NASA)

The U.S. Space Force and its antecedents have a long history of cooperation with NASA, as the lead government agencies for military and civil spaceflight. The Space Force's predecessors in the Air Force, Navy, and Army provided NASA with its early space launch vehicles and most of its astronauts.

The Space Force hosts NASA launch operations at Vandenberg Space Force Base and Cape Canaveral Space Force Station. NASA occasionally hosts U.S. Space Force heavy launches out of Kennedy Space Center. The Space Force continues to support NASA's human spaceflight missions with range support of Space Launch Delta 45 and tracks threats to the International Space Station and other crewed spacecraft.

The National Advisory Committee for Aeronautics (NACA), established on March 3, 1915, played a pivotal role in shaping aeronautical research and development in the United States. Founded during the early years of World War I, NACA was modeled after European institutions, such as the British Advisory Committee for Aeronautics, to coordinate and promote cooperation between industry, academia, and government. Its mandate was to oversee the scientific study of flight and devise practical solutions to the challenges of aviation.

NACA's early contributions to aeronautics were profound. Among its most significant developments were the NACA duct, a revolutionary air intake design used in both aviation and automotive applications, and the NACA cowling, which streamlined engine cooling for aircraft. NACA also pioneered airfoil design, creating a series of airfoils that are still widely used in aircraft manufacturing today. These innovations laid the foundation for modern aviation, demonstrating NACA's far-reaching influence in both civil and military aviation sectors.

During World War II, NACA's work proved crucial. It was often referred to as the "Force Behind Our Air Supremacy" due to its successful development of high-altitude superchargers for bombers, and the breakthrough laminar flow airfoils for the P-51 Mustang fighter plane. The Mustang's superior performance was a direct result of these advances, which contributed significantly to the Allies' aerial dominance. Additionally, NACA's work on compressibility research laid the groundwork for the Bell X-1, the first aircraft to break the sound barrier, ushering in the era of supersonic flight.

The origins of NACA can be traced back to December 1912, when President William Howard Taft formed a National Aerodynamical Laboratory Commission. However, early attempts to establish a formal body for aeronautical research met with failure. It wasn't until January 1915 that renewed efforts, spearheaded by Charles D. Walcott, Secretary of the Smithsonian Institution, gained momentum. His collaboration with legislators Senator Benjamin R. Tillman and Representative Ernest W. Roberts led to the creation of NACA. With support from Assistant Secretary of the Navy Franklin D. Roosevelt, the NACA resolution was successfully appended to the Naval Appropriations Bill, enabling the committee's formal establishment. President Woodrow Wilson signed it into law on March 3, 1915.

NACA's influence on research accelerated rapidly with the establishment of testing facilities. The Langley Memorial Aeronautical Laboratory in Hampton, Virginia became the heart of NACA's operations, home to its first wind tunnels in the 1920s. These wind tunnels allowed engineers to test new concepts and enhance designs, such as the thin airfoil theory and the NACA engine cowl, which helped improve aircraft performance. By 1938, NACA had expanded significantly, boasting 426 employees and four major research facilities, including Ames Aeronautical Laboratory and Lewis Research Center.

As World War II approached, NACA's research provided essential advancements in aircraft technology. NACA engineers solved issues with the superchargers used in Boeing B-17 bombers, allowing the planes to maintain power at high altitudes. Additionally, NACA's laminar flow airfoils made the P-51 Mustang the war's premier fighter aircraft. In the postwar years, NACA's focus shifted to supersonic and transonic flight. The development of the area rule, a crucial aerodynamic principle discovered by NACA engineer Richard Whitcomb in 1951, allowed aircraft like the Convair F-102 and F11F Tiger to break the sound barrier.

By the 1950s, NACA was leading research on supersonic aircraft, notably with the Bell X-1 program. Although the U.S. Air Force commissioned the X-1, NACA was integral to its testing and research, solving the issues of compressibility that aircraft faced as they neared the speed of sound. This work earned John Stack, NACA's head of compressibility research, the Collier Trophy, shared with Chuck Yeager and Bell Aircraft.

NACA's work on supersonic aircraft culminated in the development of the B-58 Hustler, the first U.S. bomber capable of Mach 2, a testament to the area rule's success. NACA also contributed to early spaceflight research, playing a vital role in the X-15 rocket plane program, which pushed the boundaries of manned flight into the edge of space.

In 1958, NACA was dissolved, and its facilities, personnel, and ongoing projects were transferred to the newly formed National Aeronautics and Space Administration (NASA). The legacy of NACA's 43-year tenure as the nation's premier aeronautical research agency is evident in the continued use of NACA airfoils on modern aircraft, and in the principles of supersonic and spaceflight research that laid the groundwork for NASA's future success.

The National Aeronautics and Space Administration (NASA)

The National Aeronautics and Space Administration (NASA) is an independent agency of the U.S. federal government responsible for the nation's civilian space program, as well as for aeronautics and space research. Founded on July 29, 1958, NASA was established through the National Aeronautics and Space Act, succeeding the National Advisory Committee for Aeronautics (NACA), which had served the country's aeronautical research needs since 1915. NASA's creation was driven by the necessity to give the United States' space efforts a distinctly civilian character, emphasizing peaceful applications in space science and exploration.

Since its inception, NASA has spearheaded most of the United States' space exploration efforts, overseeing projects of monumental significance. Among its early programs were Project Mercury, the first American effort to send humans into space, and Project Gemini, which laid the groundwork for future spaceflights by advancing spacecraft maneuvering, long-duration missions, and spacewalks. This foundational work culminated in the historic Apollo Moon landings from 1968 to 1972, a crowning achievement that saw NASA fulfill President John F. Kennedy's ambitious goal of landing a man on the Moon before the decade's end.

Following Apollo, NASA's focus expanded to include orbital missions, including the Skylab space station—the first American space station, which provided astronauts the opportunity to conduct long-duration missions in space. In the decades that followed, NASA developed the Space Shuttle program, an iconic symbol of reusable spacecraft technology. The Shuttle, operational from 1981 to 2011, played a crucial role in deploying satellites, conducting scientific experiments, and supporting the construction and servicing of the International Space Station (ISS).

NASA remains deeply involved in the ISS, which it operates in collaboration with international partners. Alongside the ISS, NASA's Commercial Crew Program has brought private companies into the fold, fostering the development of spacecraft capable of ferrying astronauts to low-Earth orbit. The agency also oversees the development of the Orion spacecraft and the Space Launch System (SLS), two critical elements of the Artemis program, which aims to return astronauts to the Moon and establish a sustainable human presence by the end of this decade.

In parallel with human spaceflight, NASA's science division advances knowledge in a broad array of scientific fields. The Earth Observing System, composed of satellites and instruments, enhances understanding of our planet's climate and ecosystems. NASA's heliophysics division explores the Sun and its influence on the solar system, while robotic missions, such as New Horizons and the Perseverance rover, are tasked with exploring distant bodies, from Pluto to Mars. NASA's astrophysics research, exemplified by the James Webb Space Telescope and the earlier Great Observatories, probes the deepest mysteries of the universe, from the origins of the cosmos to the formation of galaxies, stars, and planets.

NASA traces its origins to NACA, the National Advisory Committee for Aeronautics, which was established in 1915 to help the United States regain its leadership in aviation. The U.S., despite its early advances in flight, had fallen behind Europe in aviation development by the onset of World War I. To address this, Congress created the Aviation Section of the U.S. Army Signal Corps in 1914, followed by the establishment of NACA the next year. Over the next several decades, NACA would drive significant advancements in aviation, conducting research that supported both military and civilian aviation. NACA's expertise also extended into the post-World War II era, as it became interested in rocketry and supersonic flight. This led to the development and testing of experimental aircraft such as the Bell X-1, the first aircraft to break the sound barrier in 1947.

The Soviet Union's successful launch of Sputnik 1 in 1957 shocked the world and intensified the Cold War competition between the United States and the USSR, marking the start of the Space Race. In response, the U.S. government sought to consolidate its space efforts, which had been scattered across various military and civilian agencies. President Dwight D. Eisenhower's administration moved swiftly, first by creating the Advanced Research Projects Agency (ARPA) under the Department of Defense, and then establishing NASA as the country's civilian space agency. NASA began operations on October 1, 1958, absorbing NACA and its 8,000 employees, as well as key facilities such as the Jet Propulsion Laboratory (JPL) and the Army Ballistic Missile Agency, led by rocket pioneer Wernher von Braun.

The establishment of NASA marked the beginning of America's full-scale commitment to human space exploration. NASA's first human spaceflight program, Project Mercury, aimed to put an American in space and return them safely to Earth. This ambitious goal built upon earlier military efforts such as the Air Force's Man in Space Soonest project. To manage this monumental task, NASA created the Space Task Group, which would oversee the Mercury missions.

NASA's first crewed spaceflight occurred on May 5, 1961, when astronaut Alan Shepard became the first American in space aboard Freedom 7, achieving a suborbital flight. This mission came on the heels of Soviet cosmonaut Yuri Gagarin's historic first human spaceflight less than a month earlier. NASA's first orbital flight followed on February 20, 1962, when John Glenn orbited Earth three times aboard Friendship 7. Glenn's flight was not without challenges—he had to manually control the spacecraft during part of the mission due to an autopilot malfunction.

Project Mercury concluded with its sixth mission in May 1963, when astronaut Gordon Cooper orbited the Earth 22 times in Faith 7. Project Mercury was a resounding success, achieving its primary objectives of orbiting a human in space, developing tracking and control systems, and understanding the physiological and psychological effects of spaceflight on humans.

While NASA's focus shifted toward space, it maintained its commitment to cutting-edge aeronautics research. One of the most significant programs during this period was the North American X-15, a joint NASA-U.S. Air Force project. The X-15 was a rocket-powered aircraft capable of reaching hypersonic speeds, and it became the first aircraft to cross from the atmosphere into outer space, blurring the lines between aeronautics and astronautics.

The X-15 program not only advanced understanding of hypersonic flight but also served as a critical testbed for technologies that would be used in the Apollo program. The research conducted through the X-15 contributed to the development of spacecraft heat shields, high-altitude flight control systems, and advanced propulsion methods such as ramjets and scramjets.

Escalating tensions during the Cold War between the United States and the Soviet Union ignited the Space Race, with President John F. Kennedy making one of the most defining decisions of the era. In a bold challenge to Soviet space superiority, Kennedy tasked NASA with landing an American on the Moon and returning him safely to Earth by the end of the 1960s. This ambitious goal was publicly declared on May 25, 1961, during his "Urgent National Needs" speech to Congress, where he famously stated:

"I believe this Nation should commit itself to achieving the goal, before this decade is out, of landing a man on the Moon and returning him safely to Earth. No single space project in this period will be more impressive to mankind, or more important for the long-range exploration of space; and none will be so difficult or expensive to accomplish."

To ensure NASA's success, Kennedy appointed James E. Webb as the agency's administrator, providing steadfast support even amid criticism from figures such as former President Dwight Eisenhower and 1964 presidential candidate Barry Goldwater. Kennedy reinforced public support with his inspiring "We choose to go to the Moon" speech at Rice University on September 12, 1962. Despite challenges, NASA's budget grew, and it is estimated that at its peak, 5% of the American workforce was involved in some aspect of the Apollo program.

NASA modeled the Apollo program after the Department of Defense's approach to managing intercontinental ballistic missile programs, adopting a strategy of using redundant systems to ensure mission success. Major General Samuel C. Phillips, an Air Force officer, was appointed director of the Apollo program, overseeing its vast complexities. The monumental development of the Saturn V rocket, which would carry astronauts to the Moon, was led by Wernher von Braun and his team at NASA's Marshall Space Flight Center. The Apollo spacecraft itself was designed and built by North American Aviation, while the Lunar Module, the craft that would land on the Moon's surface, was constructed by Grumman.

Before the Apollo program could proceed to the Moon, NASA needed to develop the necessary spaceflight skills and technologies. This led to Project Gemini, a precursor to Apollo that focused on critical mission elements such as spacecraft maneuvering, long-duration spaceflights, and rendezvous and docking operations in space. Using modified Titan II rockets, Gemini missions demonstrated vital techniques, including fuel cell technology, spacewalks, and the ability to perform complex orbital maneuvers.

In tandem with Gemini, NASA advanced its robotic lunar exploration efforts. The Ranger program, initiated in the 1950s, aimed to capture images of the lunar surface but was plagued by early failures. However, the Lunar Orbiter and Surveyor programs, designed to map and study the Moon, were more successful. The Lunar Orbiter missions provided detailed maps of potential Apollo landing sites, while the Surveyor spacecraft performed soft landings on the Moon, analyzing the surface composition and providing crucial data on the lunar regolith.

Despite setbacks, including the tragic Apollo 1 fire in 1967, which claimed the lives of astronauts Gus Grissom, Ed White, and Roger Chaffee, NASA pressed forward. Rigorous safety improvements were made, allowing the Apollo program to continue toward its goal.

Apollo 8 became the first human-crewed mission to leave low Earth orbit and reach the Moon. Launched on December 21, 1968, the mission was commanded by Frank Borman, with James Lovell and William Anders serving as the crew. On Christmas Eve, they became the first humans to orbit the Moon, witnessing Earthrise—a view of Earth from the Moon's perspective. This mission was not only a technical triumph but also a profound moment for humanity, as the astronauts beamed back images of Earth as a fragile globe suspended in the vastness of space.

On July 20, 1969, NASA achieved the pinnacle of human space exploration. Apollo 11, commanded by Neil Armstrong and piloted by Buzz Aldrin and Michael Collins, successfully landed on the Moon. As Armstrong descended the ladder of the Lunar Module and placed his foot on the lunar surface, he proclaimed the iconic words: "That's one small step for man, one giant leap for mankind." This historic moment, broadcast worldwide, marked the fulfillment of Kennedy's goal and the United States' victory in the Space Race. Over the next few hours, Armstrong and Aldrin conducted experiments, collected lunar samples, and planted the American flag, while Collins orbited above in the Command Module.

The Apollo program would conduct five more successful lunar landings, each contributing to scientific discovery and furthering human knowledge of the Moon. Apollo 17, the final lunar mission, was completed in December 1972, marking the end of an era in space exploration.

Wernher von Braun had long advocated for the development of a space station, and after the Apollo missions concluded, NASA launched Skylab, the United States' first space station, on May 14, 1973. The station, constructed from repurposed Saturn V components, faced significant challenges. During launch, Skylab was damaged, requiring emergency repairs by the first crew, who successfully restored its functionality. Skylab hosted three crewed missions, conducting valuable scientific experiments in fields such as solar observation and human physiology in space. However, due to budget constraints and the absence of a shuttle system to boost its orbit, Skylab was deorbited in 1979.

NASA's final Apollo mission was a diplomatic milestone during the Cold War. In 1975, the Apollo-Soyuz Test Project brought together American and Soviet astronauts in the first international human spaceflight. The mission involved the docking of an Apollo spacecraft with a Soviet Soyuz capsule, symbolizing a brief thaw in the superpower rivalry and marking the final flight of an Apollo spacecraft.

During the 1960s, NASA launched a pioneering space science and interplanetary exploration program that would ultimately revolutionize humanity's understanding of the Solar System. The Jet Propulsion Laboratory (JPL) became the lead NASA center for robotic exploration, overseeing a series of successful missions aimed at studying the inner planets. The Mariner program, NASA's flagship initiative for interplanetary exploration during this period, sent probes to Venus, Mars, and Mercury, yielding significant discoveries about these worlds. Despite these early successes, congressional concerns over funding led NASA Administrator James Webb to suspend further interplanetary missions, redirecting NASA's focus and resources toward the Apollo program.

Following the successful conclusion of the Apollo Moon landings, NASA resumed its interplanetary exploration efforts with a renewed focus on space science. Venus, often considered Earth's "sister planet" due to its size and proximity, was the first planet targeted for exploration. NASA's Mariner 2 spacecraft, launched in 1962, became the first successful mission to Venus, revealing it to be an incredibly hot and inhospitable planet with surface temperatures capable of melting lead. Subsequent missions such as the Pioneer Venus project in the 1970s and the Magellan mission in the 1980s and 1990s mapped Venus' surface using radar, providing detailed insights into its geological features. Venus was later visited by various spacecraft on their way to other destinations within the Solar System.

Mars, long a subject of fascination due to its potential for harboring life, was another prime target for NASA's exploration efforts. Mariner 4 performed the first flyby of Mars in 1965, followed by Mariner 6 and Mariner 7 in 1969, and the first orbital mission to Mars, Mariner 9, in 1971. NASA's Viking program in the mid-1970s achieved the first successful landings on the Martian surface. Both Viking 1 and Viking 2 landers transmitted the first detailed images of the planet and conducted experiments in search of microbial life. While the Viking experiments did not find definitive evidence of life, they laid the foundation for future Mars missions.

After a hiatus, NASA returned to Mars in the 1990s with the Mars Global Surveyor and Mars Pathfinder missions. The latter mission introduced Sojourner, the first rover to explore the Martian surface. In the early 2000s, NASA deployed the 2001 Mars Odyssey orbiter, followed by the Spirit and Opportunity rovers in 2004, which conducted extended missions far exceeding their planned lifespans. The Mars Reconnaissance Orbiter (2005) and Phoenix lander (2007) further expanded humanity's knowledge of Mars, with Phoenix conducting detailed analyses of the Martian soil and ice.

In 2012, the Curiosity rover landed on Mars, discovering evidence that the planet once had environmental conditions favorable for microbial life, including the presence of key chemical ingredients for life. Curiosity also found that the radiation levels on Mars were comparable to those on the International Space Station, enhancing the feasibility of future human exploration. In subsequent years, NASA launched the Mars Atmosphere and Volatile Evolution (MAVEN) mission to study the planet's atmosphere and the InSight lander (2018) to investigate the planet's interior. In 2021, the Perseverance rover arrived on Mars, accompanied by the first extraplanetary helicopter, Ingenuity, which successfully demonstrated powered flight on another world.

NASA's interplanetary missions extended beyond Venus and Mars. In 2004, the MESSENGER spacecraft was launched to explore Mercury, becoming the first spacecraft to orbit the innermost planet. NASA also turned its attention to the outer Solar System, beginning with the Pioneer program in the 1970s. Pioneer 10 became the first spacecraft to fly by Jupiter, while Pioneer 11 provided the first close-up images of Saturn. Both probes later became the first human-made objects to leave the Solar System.

The Voyager program, launched in 1977, achieved some of the most remarkable milestones in space exploration. Voyager 1 and Voyager 2 conducted flybys of Jupiter, Saturn, Uranus, and Neptune, capturing unprecedented images and data before continuing their journeys into interstellar space. The Galileo spacecraft, deployed from the Space Shuttle in 1989, became the first to orbit Jupiter, discovering evidence of subsurface oceans on its moon Europa and observing volcanic activity on Io.

In 1997, NASA, in collaboration with the European Space Agency and the Italian Space Agency, launched the Cassini–Huygens mission to Saturn. The mission provided groundbreaking insights into Saturn's moons, particularly Titan and Enceladus, both of which exhibited conditions that may be conducive to life. The Huygens probe landed on Titan, revealing a landscape shaped by liquid methane lakes, while Cassini discovered geysers of water ice erupting from Enceladus' south pole, suggesting the presence of a subsurface ocean.

NASA's outer Solar System exploration culminated in the New Horizons mission, launched in 2006, which became the first spacecraft to visit Pluto in 2015. New Horizons provided humanity with its first close-up images of the distant dwarf planet and continued its journey through the Kuiper Belt, studying other icy bodies on the fringes of the Solar System.

In addition to interplanetary exploration, NASA has launched numerous space telescopes to study the universe. In the 1960s, the Orbiting Astronomical Observatories were among NASA's first space telescopes, providing observations in ultraviolet, gamma-ray, x-ray, and infrared spectra. The Great Observatories program, launched in the 1990s and early 2000s, further advanced space-based astronomy. The Hubble Space Telescope, launched in 1990, revolutionized humanity's view of the cosmos, capturing images of galaxies billions of light-years away. Although initially plagued by a flaw in its mirror, Hubble's capabilities were restored through a series of Space Shuttle servicing missions.

Other Great Observatories included the Compton Gamma Ray Observatory, launched in 1991, which provided insights into the most energetic phenomena in the universe, and the Chandra X-ray Observatory, launched in 1999, which observed black holes, quasars, and dark matter. The Spitzer Space Telescope, launched in 2003, studied the universe in infrared light, revealing the presence of brown dwarf stars and other celestial objects hidden from visible light telescopes.

NASA's space telescopes have also contributed to the study of cosmology and the origins of the universe. The Cosmic Background Explorer (COBE) and the Wilkinson Microwave Anisotropy Probe (WMAP) provided critical evidence supporting the Big Bang theory. More recently, the James Webb Space Telescope, launched in 2021, was designed to observe the first galaxies that formed after the Big Bang, continuing NASA's legacy of groundbreaking astronomical discoveries.

NASA's telescopes have not only explored the distant universe but also contributed to the search for habitable planets beyond our Solar System. The Kepler space telescope, launched in 2009, identified thousands of exoplanets, including Kepler-22b, a planet located in the habitable zone of its star, where conditions may be right for liquid water and life.

Closer to home, NASA has launched a series of Earth observation satellites to study the planet's climate and environment. The Television Infrared Observation Satellite (TIROS), launched in 1960, became the world's first weather satellite. Subsequent missions, including the Nimbus and Landsat programs, provided detailed observations of Earth's atmosphere, land use, and ocean conditions. NASA's Earth science efforts, in collaboration with the National Oceanic and Atmospheric Administration (NOAA), have contributed to critical discoveries, such as ozone depletion and the effects of climate change.

NASA's pursuit of a reusable spaceplane dates back to the 1960s, when it envisioned such a vehicle as part of a broader program aimed at providing routine, cost-effective logistical support to a space station orbiting Earth. This space station would serve as a hub for future lunar and Mars missions. A reusable launch vehicle, such as a spaceplane, would have eliminated the need for expensive, expendable boosters like the Saturn V, which were pivotal during the Apollo program.

By 1969, NASA designated the Johnson Space Center as the lead center for designing, developing, and manufacturing the Space Shuttle orbiter, while the Marshall Space Flight Center focused on the launch system. Development efforts were influenced by NASA's previous work on lifting body aircraft, notably the joint NASA-U.S. Air Force X-24 program. These early efforts directly contributed to the Space Shuttle's design and other future hypersonic aircraft concepts.

In 1972, NASA officially began developing the Space Shuttle. The orbiter design and engines were awarded to Rockwell International, the external fuel tank to Martin Marietta, and the solid rocket boosters to Morton Thiokol. NASA eventually acquired six orbiters: Enterprise, Columbia, Challenger, Discovery, Atlantis, and Endeavour.

The Space Shuttle program marked a new era for NASA's Astronaut Corps. While previous astronaut selections were dominated by military test pilots, the Shuttle allowed NASA to recruit scientists, engineers, and medical professionals. Sally Ride, who became the first American woman to fly in space on the STS-7 mission, exemplified this shift. Additionally, the program opened opportunities for international exchange astronauts, reflecting NASA's growing partnerships with allied nations.

The first flight of the Space Shuttle occurred in 1981 with the launch of Columbia on the STS-1 mission. This mission tested the spacecraft's capabilities, marking the beginning of NASA's plan to replace expendable rockets like the Atlas, Delta, and Titan series, as well as the European Space Agency's Ariane rocket. With the introduction of Spacelab, developed by the European Space Agency, Shuttle missions gained unprecedented scientific capabilities, expanding NASA's research potential beyond previous missions.

The Shuttle program achieved numerous milestones. In 1982, on the STS-5 mission, NASA launched its first commercial satellites. Then, in 1984, the STS-41-C mission aboard Challenger conducted the first on-orbit satellite servicing mission, capturing and repairing the malfunctioning Solar Maximum Mission satellite. The Shuttle also demonstrated the ability to return malfunctioning satellites to Earth, such as the Palapa B2 and Westar 6 satellites, which were repaired and relaunched.

Despite these successes, the Shuttle program faced criticism for not being as reusable or cost-effective as initially promised. The Challenger disaster in 1986, during the STS-51L mission, resulted in the loss of all seven astronauts and grounded the Shuttle fleet for nearly three years. The disaster forced NASA to modify the Shuttle's design and operations, resulting in improved safety measures when the Shuttle returned to flight in 1988 with the STS-26 mission.

In the 1990s, following the end of the Cold War and the dissolution of the Soviet Union, the U.S. and Russia initiated the Shuttle-Mir program, marking a new era of international collaboration in space. The first Russian cosmonaut flew aboard the STS-60 mission in 1994, and NASA's Space Shuttle began rendezvousing with the Russian space station Mir. In 1995, the Atlantis orbiter docked with Mir on the STS-71 mission, fulfilling one of the Space Shuttle's original objectives of station resupply and crew transfer. The Shuttle-Mir program continued until 1998, when operational accidents aboard Mir concluded the partnership.

Tragedy struck again in 2003 when the Columbia orbiter was lost during reentry on the STS-107 mission, resulting in the deaths of seven astronauts. The Columbia disaster accelerated plans to retire the Shuttle program. President George W. Bush directed that once the International Space Station (ISS) was complete, the Shuttle would be decommissioned. Following a series of final missions, including servicing the Hubble Space Telescope, the Shuttle program officially ended with the STS-135 mission in 2011, which resupplied the ISS.

NASA's vision of a permanent space station did not end with Skylab's reentry in 1979. The agency immediately began advocating for a larger, more advanced space station to support long-term human habitation, scientific research, and deep space exploration. President Ronald Reagan became a key supporter of this vision, declaring in his 1984 State of the Union address that the United States would build a permanently manned space station within a decade.

NASA's proposal, named Space Station Freedom, aimed to be an international project, with contributions from partner nations to solidify its global importance. In 1985, the United States signed an agreement with thirteen countries, including members of the European Space Agency, Canada, and Japan, to collaborate on its development. However, budget concerns and multiple redesigns plagued the project in the early 1990s. Political pressure to reduce costs intensified, and President Bill Clinton's administration ultimately restructured the program.

In 1993, the Clinton administration announced a landmark agreement with the Russian Federation to transform Space Station Freedom into the International Space Station (ISS). This collaboration revitalized Russia's space program while reducing costs for the U.S. NASA, along with its international partners, constructed and launched various ISS modules, turning the station into a symbol of global cooperation in space exploration. Despite NASA's original budget estimate of $17.4 billion, the cost of the ISS eventually grew to over $150 billion, with the U.S. contributing two-thirds of the total.

The completion of the ISS was accelerated following the Columbia disaster, which grounded the Shuttle fleet and left NASA dependent on Russian Soyuz spacecraft to transport astronauts to the station. Even after the Shuttle's retirement in 2011, the ISS continued to operate as a hub for international research and a platform for advancing technologies needed for future missions to the Moon, Mars, and beyond.

In the 1980s, NASA explored next-generation spaceplanes through a joint program with the Department of Defense to develop the Rockwell X-30 National Aerospace Plane. Intended as a single-stage-to-orbit vehicle, the X-30 aimed to address the limitations of the Shuttle, offering a fully reusable system with both civilian and military applications. However, with the end of the Cold War, budget constraints led to the program's cancellation in 1992, before the X-30 could achieve flight status.

Following the tragic loss of the Space Shuttle Columbia in 2003, President George W. Bush initiated the Constellation program to ensure the smooth replacement of the Space Shuttle and reignite human space exploration beyond low Earth orbit. The program aimed to return astronauts to the Moon using technologies and equipment inherited from the Space Shuttle era. However, in 2010, the Obama Administration canceled Constellation, citing delays and budget overruns. This decision prompted a letter from former astronauts Neil Armstrong, Gene Cernan, and Jim Lovell, who cautioned that without a new human spaceflight capability, the United States risked becoming a second- or third-rate space power.

Efforts to involve the private sector in space exploration had been discussed as early as the Reagan Administration. During the 1990s, NASA partnered with Lockheed Martin to develop the Lockheed Martin X-33, a demonstrator for the VentureStar spaceplane. VentureStar was intended to replace the Space Shuttle and provide a fully reusable launch vehicle. However, the project faced significant technical challenges and was canceled in 2001. Despite its failure, the program marked a turning point, as it was the first time a commercial company had invested substantial resources in spacecraft development. Around this time, space tourism also began to emerge, with Dennis Tito becoming the world's first space tourist in 2001. Tito flew to the International Space Station aboard a Russian spacecraft, despite NASA's opposition to private individuals in space.

The idea of commercializing space exploration gained momentum with advocates like former astronaut Buzz Aldrin. Aldrin argued that allowing private companies to operate space systems would enable NASA to refocus on its core mission of research, development, and deep space exploration. Under this vision, NASA could direct its resources toward returning humans to the Moon and eventually sending astronauts to Mars, while private companies managed routine orbital operations. NASA embraced this shift in strategy through its Commercial Crew Program, which initially contracted private companies for cargo delivery to the International Space Station.

In 2020, NASA marked a major milestone when SpaceX's Crew-1 mission launched American astronauts from U.S. soil for the first time since the retirement of the Space Shuttle in 2011. This mission, which flew under NASA's Commercial Crew Program, also ended the agency's reliance on Russian spacecraft for crewed missions. SpaceX had already achieved numerous cargo deliveries to the ISS, but Crew-1 was the first operational mission to carry astronauts, demonstrating the viability of NASA's partnership with the private sector.

NASA's focus on deep space exploration led to the announcement of the Artemis program in 2019. Artemis aims to return humans to the Moon and establish a permanent presence on the lunar surface, laying the groundwork for future missions to Mars. The program's objectives include the construction of a lunar base, developing sustainable exploration capabilities, and conducting critical scientific research. To support international cooperation and commercial opportunities, NASA introduced the Artemis Accords, a set of principles governing behavior and activities on the Moon, in collaboration with partner nations.

As NASA pushed forward with its vision for the future, it created the Moon to Mars Program office in 2023. This office is responsible for overseeing all lunar and Mars exploration projects, managing mission architectures, and establishing timelines for these ambitious goals. Through this framework, NASA aims to advance human space exploration into the solar system, with the Moon serving as a key stepping stone for future missions to Mars. The Artemis program, paired with growing commercial partnerships, represents a new era in space exploration, driven by both governmental and private sector efforts to reach beyond Earth's orbit.

The International Space Station (ISS) is the culmination of multiple space station projects, most notably NASA's Space Station Freedom, Russia's Mir-2, Europe's Columbus, and Japan's Kibō laboratory module. Originally, NASA had intended to develop Space Station Freedom independently in the 1980s, but budget constraints led to a merger of these international projects in 1993, creating a multinational effort managed by NASA, the Russian Federal Space Agency (RKA), the Japan Aerospace Exploration Agency (JAXA), the European Space Agency (ESA), and the Canadian Space Agency (CSA).

The ISS is composed of pressurized modules, external trusses, solar arrays, and other components built in factories worldwide. These components were launched using Russian Proton and Soyuz rockets, as well as the American Space Shuttle, beginning in 1998. The U.S. Orbital Segment was completed in 2009, followed by the Russian Orbital Segment in 2010. The station operates under intergovernmental agreements that divide it into two distinct areas: the Russian Orbital Segment, owned by Russia (except for the Zarya module), and the U.S. Orbital Segment, shared among the other international partners.

Long-duration missions on the ISS are referred to as ISS Expeditions, with crew members typically spending about six months aboard the station. Initially, expedition crews consisted of three members, but this was temporarily reduced to two following the Columbia disaster. Since 2009, expedition crews generally consist of six members, although recent Commercial Crew Program missions typically carry four astronauts. The ISS has been continuously inhabited for over 23 years, surpassing the record previously held by Russia's Mir station. Over time, it has hosted astronauts and cosmonauts from 15 different nations.

The ISS is the largest artificial satellite in Earth's orbit, visible to the naked eye from the ground. As of 2024, it remains the most massive and voluminous space station ever constructed. Astronauts travel to and from the ISS aboard Russian Soyuz spacecraft, SpaceX's Dragon, and Boeing's Starliner capsules. Additionally, uncrewed cargo spacecraft—including the Russian Progress, European Automated Transfer Vehicle (ATV), Japanese H-II Transfer Vehicle (HTV), SpaceX's Dragon, and Northrop Grumman's Cygnus—deliver supplies to the ISS. Prior to its retirement in 2011, the Space Shuttle also transported cargo and crew members to the station. Following the Shuttle's retirement, NASA relied on the Russian Soyuz for crew transport until SpaceX began crewed flights in 2020.

The ISS is expected to remain operational until 2030, after which it will be decommissioned and deorbited in a controlled descent into Earth's atmosphere.

NASA's Commercial Resupply Services (CRS) program contracts private companies to deliver cargo to the ISS. The CRS program began in 2008, awarding $1.6 billion to SpaceX for twelve missions using its Dragon spacecraft, and $1.9 billion to Orbital Sciences (now Northrop Grumman) for eight Cygnus missions, covering deliveries through 2016. Both companies developed their own launch vehicles—SpaceX with the Falcon 9 and Orbital with the Antares rocket—to support these missions.

SpaceX conducted its first operational resupply mission, CRS-1, in 2012, followed by Orbital's Cygnus CRS Orb-1 mission in 2014. In 2015, NASA extended the CRS-1 contracts, awarding SpaceX twenty flights and Orbital twelve flights. A second phase of contracts, CRS-2, was solicited in 2014, and in January 2016, NASA awarded contracts to SpaceX, Northrop Grumman (formerly Orbital), and Sierra Nevada Corporation's Dream Chaser spacecraft, for missions beginning in 2019 and expected to continue through 2024. In 2022, NASA extended the CRS-2 contracts, awarding six additional missions each to SpaceX and Northrop Grumman.

Recent milestones include the February 2022 delivery of Cygnus NG-17 by Northrop Grumman and SpaceX's 25th CRS mission in July 2022. The Dream Chaser spacecraft is scheduled to debut with its first cargo delivery to the ISS in 2024.

The Commercial Crew Program (CCP) is NASA's effort to contract private companies to transport astronauts to and from the ISS, marking a significant shift in how human spaceflight is conducted. SpaceX began providing crew transportation services in 2020 with its Crew Dragon spacecraft, and Boeing's Starliner spacecraft is set to begin operational missions in 2024. Under the CCP, NASA has contracted six operational missions from Boeing and fourteen from SpaceX, ensuring sufficient crew support for the ISS through 2030.

The spacecraft used in the CCP are owned and operated by the respective private companies, with NASA purchasing crew transportation as a commercial service. Each mission typically carries four astronauts, with the option for a fifth passenger if needed. The spacecraft remain docked to the ISS for the duration of each six-month mission, and overlapping missions ensure that crew rotations are seamless.

SpaceX's Crew Dragon is launched atop the Falcon 9 Block 5 rocket and returns to Earth via ocean splashdown near Florida. The first operational CCP mission, SpaceX Crew-1, launched on November 16, 2020, marking the return of American crewed spaceflight from U.S. soil after nearly a decade of reliance on the Russian Soyuz. Boeing's Starliner will be launched aboard the Atlas V rocket and will return to Earth on land, using airbags to cushion its landing at designated sites in the western United States.

As NASA continues its partnership with commercial space companies, the CCP ensures that human access to the ISS remains uninterrupted, while allowing NASA to focus on future deep space exploration missions.

Since 2017, NASA's primary crewed spaceflight program has been the Artemis program, which aims to return humans to the Moon and eventually pave the way for missions to Mars. This ambitious initiative relies on partnerships with U.S. commercial space companies and international collaborators, including the European Space Agency (ESA), Japan Aerospace Exploration Agency (JAXA), and Canadian Space Agency (CSA). The program's goal is to land "the first woman and the next man" on the lunar south pole by 2025, marking the first crewed lunar landing since 1972. Artemis represents the first step toward establishing a sustainable human presence on the Moon, with the long-term vision of supporting a lunar economy and laying the foundation for human missions to Mars.

The Orion Crew Exploration Vehicle, initially developed under the canceled Constellation program, was repurposed for the Artemis missions. The program's first uncrewed mission, Artemis 1, successfully launched on November 16, 2022, sending an Orion spacecraft on a distant retrograde orbit around the Moon before returning safely to Earth. This mission served as a crucial test of NASA's Space Launch System (SLS) and the Orion spacecraft, demonstrating the system's capability for deep space exploration.

Artemis 2, scheduled for 2025, will be the program's first crewed mission, sending four astronauts on a 10-day lunar flyby. This mission will mark NASA's return to crewed lunar exploration for the first time in over 50 years. Following this, Artemis 3, planned for no earlier than September 2026, aims to land astronauts on the Moon's surface, including the first woman and next man, in the lunar south pole region. The Artemis missions will gradually increase in complexity, culminating in the long-term goal of sustainable lunar operations.

In support of these missions, NASA launched the Commercial Lunar Payload Services (CLPS) program, contracting private companies such as Intuitive Machines, Firefly Space Systems, and Astrobotic to develop and land robotic probes on the lunar surface. These probes will deliver scientific instruments and technology demonstrations, providing vital information to prepare for human exploration.

On April 16, 2021, NASA selected SpaceX's Lunar Starship as the Human Landing System (HLS) for the Artemis program. For future lunar missions, the SLS will launch four astronauts aboard the Orion spacecraft, which will travel to lunar orbit. There, the crew will transfer to SpaceX's Starship for the final descent to the lunar surface. The Starship will play a key role in NASA's plans for sustained lunar exploration.

The original goal of landing astronauts on the Moon by 2024 was delayed due to technical challenges, budget constraints, and the impacts of the COVID-19 pandemic, pushing the target date to 2025. Additional Artemis missions, Artemis 4, Artemis 5, and Artemis 6, are planned to take place between 2028 and 2031.

A critical component of the Artemis program is the construction of the Lunar Gateway, a small space station that will orbit the Moon. The Gateway will serve as a staging point for lunar landings and scientific research. Its construction is expected to begin in 2027, with the launch of the first two modules: the Power and Propulsion Element (PPE) and the Habitation and Logistics Outpost (HALO). Operations aboard the Gateway will commence with Artemis 4, which will deliver a crew of four astronauts to the station in 2028.

NASA's long-term vision extends beyond the Moon. The NASA Transition Authorization Act of 2017 directed the agency to develop the capability to send humans to Mars orbit, or potentially to the Martian surface, by the 2030s. The Artemis program, along with its international and commercial partnerships, represents a crucial stepping stone toward achieving this objective.

In parallel with the Artemis program, NASA is supporting the development of commercial space stations through its Commercial Low Earth Orbit (LEO) Destinations program. This initiative aims to foster the construction of private space stations by the end of the decade to replace the ISS, which is scheduled for retirement in 2030. NASA has selected three companies to spearhead this effort: Blue Origin, with its Orbital Reef station concept; Nanoracks, with the Starlab station; and Northrop Grumman, which is developing a station based on the HALO module for the Lunar Gateway. These private space stations will ensure continued U.S. presence in low Earth orbit, supporting research, commercial ventures, and international collaboration.

NASA has a long and successful history of robotic space exploration, with over 1,000 uncrewed missions designed to explore Earth, the Solar System, and beyond. These missions provide critical data, enabling scientists to study distant planets, moons, asteroids, and other celestial bodies.

The process of selecting robotic missions follows a well-defined mission development framework. This system allows NASA to assess cost, schedule, and technical risks, enabling competitive selection of mission proposals. Proposals are developed by teams of principal investigators from across NASA, the U.S. government, and the private sector, fostering innovation and collaboration. NASA's robotic exploration missions are categorized under several key programs.

The Explorer Program traces its roots back to the dawn of the U.S. space program. Today, it consists of three mission classes: Small Explorers (SMEX), Medium Explorers (MIDEX), and University-Class Explorers (UNEX). These missions focus on heliophysics and astrophysics research and provide frequent, cost-effective opportunities for innovative scientific investigations. SMEX missions are capped at $150 million (2022 dollars), while MIDEX missions typically have cost caps of $350 million. The Explorer program is managed by NASA's Goddard Space Flight Center.

NASA's Discovery Program supports the development of low-cost, high-impact robotic missions that address specific scientific goals in planetary exploration. Under the program, scientists and engineers are invited to propose mission concepts, which then undergo a competitive selection process. Recent Discovery missions have operated under a cost cap of $500 million. NASA's Marshall Space Flight Center manages the program.

In 2021, NASA announced that two new Discovery missions, DAVINCI+ and VERITAS, were selected to explore Venus. These missions, scheduled to launch between 2028 and 2030, aim to study Venus' atmosphere and surface, providing key insights into the planet's geology and climate. The selection process for these missions followed a competitive evaluation, where they beat out competing proposals to explore Jupiter's moon Io and Neptune's moon Triton. Both missions will advance humanity's understanding of the inner workings of Venus, while Discovery continues to foster new exploration opportunities across the Solar System.

The New Frontiers Program was initiated by NASA to pursue key goals in Solar System exploration, focusing on high-priority objectives identified by the planetary science community. Its primary aim is to utilize medium-class spacecraft to conduct scientific investigations that yield significant returns. This program builds upon the framework established by the earlier Discovery Program, but with greater flexibility, allowing for higher cost caps and longer mission timelines. Recent missions have operated under a defined budget of approximately $1 billion per mission, contributing to a slower cadence of new opportunities, with missions launched every few years. Two prominent examples of New Frontiers missions are OSIRIS-REx, which studied asteroid Bennu and returned samples to Earth, and New Horizons, which provided the first close-up images of Pluto and ventured into the Kuiper Belt.

Looking ahead, NASA has announced that the next opportunity for proposing missions under the New Frontiers Program will be no later than the fall of 2024. This initiative continues to shape NASA's strategy of exploring the Solar System with medium-class spacecraft, pushing the boundaries of planetary science with each new mission.

In contrast to the New Frontiers Program, Large Strategic Missions (formerly known as Flagship missions) are high-budget, complex projects that often span multiple NASA centers and involve international collaboration. These missions, including the James Webb Space Telescope (JWST), are monumental efforts designed to address major scientific objectives. JWST, which took over 20 years to develop, represents one of NASA's most significant strategic investments. Launched in 2021, it now operates in a halo orbit around the Sun-Earth L2 point, observing distant galaxies and unraveling mysteries of the universe with unparalleled sensitivity in the infrared spectrum. Other iconic strategic missions include the Voyager probes, launched in the 1970s, which are still transmitting data as they journey into interstellar space, and the Hubble Space Telescope, whose service missions by the Space Shuttle extended its operational life and enabled groundbreaking astronomical discoveries.

NASA's Planetary Science Missions continue to explore diverse celestial bodies within the Solar System. From the Lunar Reconnaissance Orbiter orbiting the Moon to the Perseverance Rover traversing Mars, NASA remains at the forefront of interplanetary exploration. The Juno spacecraft, currently orbiting Jupiter, has provided unprecedented data on the planet's atmosphere and magnetic field, while its extended mission will include flybys of the Jovian moon Io in 2023 and 2024. Similarly, the New Horizons mission, after its historic flyby of Pluto in 2015, continues its exploration of Kuiper Belt Objects, pushing humanity's knowledge of the outer Solar System.

One of NASA's most ambitious planetary missions, the Mars Science Laboratory, successfully landed the Curiosity Rover on Mars in 2012. Curiosity has since been investigating the planet's surface, searching for signs of past habitability. Additionally, the MAVEN spacecraft, which entered Mars orbit in 2014, continues to study the planet's atmosphere, shedding light on how Mars lost its water and atmosphere over time. Upcoming missions, such as the Europa Clipper, planned for launch in 2024, will explore Jupiter's icy moon Europa, believed to harbor a subsurface ocean that may hold the potential for life. Similarly, the Dragonfly mission, slated for a 2027 launch, will send a rotorcraft to explore Saturn's moon Titan, offering a unique perspective on its complex atmosphere and surface.

NASA's Astrophysics Missions are managed by the agency's Science Mission Directorate, focusing on observing and understanding the universe across a broad spectrum of wavelengths. The Great Observatories, a series of space telescopes launched in the 1980s and 1990s, have contributed significantly to this effort. The Hubble Space Telescope, launched in 1990, continues to operate, providing invaluable images and data, thanks to servicing missions by the Space Shuttle. The Chandra X-ray Observatory, launched in 1999, remains a powerful tool for studying high-energy regions of the universe, such as black holes and supernova remnants.

More recent missions, such as the Imaging X-ray Polarimetry Explorer (IXPE), launched in December 2021, are designed to study the polarization of X-rays emitted by exotic objects like neutron stars and black holes. This mission, a collaboration between NASA and the Italian Space Agency (ASI), falls under NASA's Small Explorers (SMEX) Program, which funds lower-cost missions focused on astrophysics and heliophysics. Another key observatory, the Neil Gehrels Swift Observatory, launched in 2004, continues to monitor gamma-ray bursts, offering insights into the most energetic events in the universe. Meanwhile, the Fermi Gamma-ray Space Telescope, launched in 2008, performs gamma-ray astronomy, providing critical data on cosmic phenomena.

The James Webb Space Telescope, launched in December 2021, represents a leap forward in astrophysics, offering unprecedented imaging resolution and sensitivity in the infrared spectrum. This allows scientists to peer deeper into the cosmos, observing the formation of galaxies, stars, and planetary systems in distant, faint regions of space, well beyond the capabilities of its predecessors like Hubble. JWST's mission is poised to revolutionize our understanding of the universe, much as the Great Observatories did in their time.

The NASA Earth Science Program, established in 1965, serves as a comprehensive initiative aimed at understanding Earth's complex systems and their response to both natural and human-induced changes. This program has evolved over several decades, incorporating a vast array of terrestrial and space-based collection systems designed to improve predictions of weather, climate, and environmental changes.

A variety of spacecraft have been launched as part of the Earth Science Program, each with specific goals. Notable missions include Aqua and Aura, which study Earth's water cycle and atmospheric composition, respectively, and the Orbiting Carbon Observatory 2 (OCO-2), which monitors carbon dioxide levels in Earth's atmosphere. Other significant missions include the Gravity Recovery and Climate Experiment Follow-on

(GRACE-FO), which tracks changes in Earth's gravity to study water movement, and ICESat-2, which measures ice sheet elevation to monitor changes in polar regions. These satellite missions are critical for developing a deeper understanding of Earth's climate and environmental dynamics.

In addition to ongoing missions, NASA is actively developing the next generation of Earth Observing Systems. These new systems are designed to address key challenges such as climate change, natural hazards, forest fires, and agricultural monitoring in real time. The most recent addition to this fleet is the GOES-T satellite (designated GOES-18 after launch in March 2022), which joins a network of U.S. geostationary weather monitoring satellites, providing enhanced capabilities for real-time weather forecasting.

NASA's Earth Science Data Systems (ESDS) program plays a vital role in managing the data collected from these missions. The ESDS ensures that data is processed, distributed, and made available for research, applications, and decision-making. This program maximizes the scientific return from NASA's missions, providing critical information to the scientific community, policymakers, and society as a whole.

The NASA Deep Space Network (DSN), established in 1963, serves as the primary communication link for NASA's interplanetary spacecraft. With ground stations strategically located near Barstow, California; Madrid, Spain; and Canberra, Australia, the DSN provides continuous communication with spacecraft across the Solar System, regardless of Earth's rotation. This global network, managed by the Jet Propulsion Laboratory (JPL), ensures that NASA maintains contact with up to 40 spacecraft simultaneously, supporting missions ranging from Mars rovers to deep space probes like Voyager.

For satellites and spacecraft closer to home, the Near Space Network (NSN) provides essential telemetry, tracking, and data communication services. Formed in 1983, the NSN combines ground-based tracking systems with the Tracking and Data Relay Satellite System (TDRS), which operates in geosynchronous orbit, providing continuous communication for low Earth orbit missions. The network, managed by Goddard Space Flight Center, supports an average of 120 to 150 daily spacecraft contacts and is essential for missions requiring real-time data, including launches and near-Earth satellites.

Another long-standing component of NASA's operations is the Sounding Rocket Program (NSRP), in place since 1959. This program, based at Wallops Flight Facility, provides the capability to launch suborbital missions, which gather data from Earth's atmosphere and space. Managed by Goddard Space Flight Center, the NSRP conducts around 20 launches annually, deploying sounding rockets from various global locations. The program supports scientific research in Earth science, heliophysics, and astrophysics, offering unique insights into phenomena that occur at specific altitudes or locations. In June 2022, NASA launched its first sounding rocket from a commercial spaceport outside the U.S., sending a Black Brant IX rocket from Arnhem Space Centre in Australia, further expanding its global launch capabilities.

The NASA Launch Services Program (LSP), which began in 1990, oversees the procurement and integration of commercial launch services for NASA's uncrewed scientific and applications missions. Based at Kennedy Space Center, LSP ensures that NASA's scientific payloads reach orbit safely and efficiently, using expendable launch vehicles purchased from commercial providers. These vehicles are capable of reaching a wide range of orbital inclinations and altitudes, making them suitable for Earth observation and interplanetary missions alike.

NASA also maintains a robust commitment to aeronautics research through its Aeronautics Research Mission Directorate (ARMD). Established as one of NASA's five mission directorates, ARMD conducts cutting-edge research in aviation, benefiting commercial, military, and general aviation sectors. This research is carried out across four major NASA centers: Ames Research Center and Armstrong Flight Research Center in California, Glenn Research Center in Ohio, and Langley Research Center in Virginia. ARMD's innovations in aeronautics have made air travel safer, more efficient, and more environmentally friendly, continuing NASA's legacy of contributing to advancements in aviation technology.

The NASA X-57 Maxwell, developed beginning in 2016, represents a pioneering effort to demonstrate the technologies necessary for a highly efficient all-electric aircraft. The primary goal of this experimental aircraft is to develop electric propulsion solutions that meet the requirements for airworthiness certification, a crucial step toward integrating electric aircraft into the commercial aviation sector. The X-57 program involves a phased development approach, with each stage—referred to as modifications—incrementally enhancing the system's capabilities. By mid-2022, the X-57 had completed ground testing and was scheduled to undertake its first flight by the end of the year. The development team comprises experts from NASA Armstrong, NASA Glenn, and NASA Langley, in collaboration with industry partners from the U.S. and Italy. This project reflects NASA's broader commitment to advancing sustainable aviation technologies.

In tandem with the X-57 program, NASA has also been a key collaborator in the Next Generation Air Transportation System (NextGen), which began in 2007 as a partnership with the Federal Aviation Administration (FAA) and industry stakeholders. The primary objective of NextGen is to modernize the National Airspace System (NAS), with a targeted completion of major advancements by 2025. This modernization is intended to increase the safety, efficiency, capacity, and flexibility of the NAS while minimizing aviation's environmental impact. At the heart of NASA's contributions is the Aviation Systems Division at NASA Ames, which operates the North Texas Research Station, a joint NASA/FAA facility. This center supports NextGen research through all stages, from concept development to prototype evaluations. NASA has played a vital role in developing advanced air traffic management tools, providing air traffic controllers and pilots with more accurate real-time information about weather, traffic flow, and routing. These innovations, supported by NASA's advanced airspace modeling tools, have already been implemented by the FAA, leading to more optimized airspace designs and improved traffic flow management.

Technology Research is another cornerstone of NASA's mission, particularly in the area of in-space nuclear power and propulsion. One key technology is the Multi-Mission Radioisotope Thermoelectric Generator (MMRTG), which converts heat from decaying plutonium-238 into electricity to power deep space missions. However, shortages of plutonium-238 have constrained NASA's deep space exploration since the early 2000s. One notable mission that was not pursued due to this shortage was New Horizons 2, a planned follow-up to the successful New Horizons mission.

NASA has since renewed its focus on nuclear propulsion technology. In July 2021, the agency awarded contracts to three companies to develop nuclear thermal propulsion (NTP) reactors, which use nuclear fission to heat a propellant and produce thrust, providing more efficient deep space travel. In early 2023, NASA announced a partnership with the Defense Advanced Research Projects Agency (DARPA) on the Demonstration Rocket for Agile Cislunar Operations (DRACO) program, aimed at demonstrating nuclear thermal propulsion in space, a key capability for future Mars missions. Later that year, NASA and DARPA awarded a $499 million contract to Lockheed Martin to design and build an experimental NTP rocket, with a planned launch in 2027.

Other forward-looking NASA initiatives include exploring innovative technologies such as Free Space Optics (FSO) for satellite communication, which utilizes laser-based optical stations to enhance satellite data transfer capabilities. NASA also seeks to harness lunar resources, including extracting water from lunar soil. In July 2020, NASA invited American universities to propose new technologies for lunar water extraction and power systems, contributing to NASA's vision of sustainable Moon exploration.

In a unique endeavor, the U.S. government tasked NASA in 2024 with creating a standardized Time System for the Moon. Named Coordinated Lunar Time, this time standard is expected to be finalized by 2026, providing a consistent and reliable timekeeping system for future lunar missions and activities.

NASA's Human Research Program (HRP) is a cornerstone of efforts to understand and mitigate the health risks of space travel. Since its inception, HRP has focused on studying the physiological effects of space on the human body and developing countermeasures to ensure the safety and well-being of astronauts during long-duration missions. Although short-term missions in low Earth orbit (LEO) and lunar exploration pose relatively minor health challenges, extended travel to Mars and other deep space destinations introduces significant medical concerns. These include bone density loss, increased radiation exposure, vision impairment, circadian rhythm disturbances, cardiovascular remodeling, and immune system changes.

In response to these challenges, HRP aims to develop portable, lightweight medical equipment capable of monitoring astronauts' health in space environments. A major milestone in this effort occurred on May 13, 2022, when SpaceX Crew-4 astronaut Samantha Cristoforetti tested NASA's rHEALTH ONE, a universal biomedical analyzer, aboard the International Space Station (ISS). The device successfully identified and analyzed biomarkers, cells, microorganisms, and proteins in real-time, marking a significant advancement in spaceflight medical diagnostics. The HRP continues to refine these technologies as NASA prepares for future missions to Mars and beyond.

In 2016, NASA established the Planetary Defense Coordination Office (PDCO) to track and assess the risks posed by near-Earth objects (NEOs), including asteroids and comets, that could potentially impact Earth. The PDCO's mission is to provide timely and accurate information about close approaches of potentially hazardous objects (PHOs) and to develop strategies to mitigate these threats. Operating under NASA's Science Mission Directorate's Planetary Science Division, the PDCO coordinates efforts with international space agencies and scientific communities to catalog and monitor NEOs.

This office expanded upon earlier efforts, notably the Spaceguard program initiated in 1998, in which the United States and international partners began systematic observations of the sky to detect NEOs. The United States Congress formally tasked NASA in 1998 with detecting 90% of NEOs larger than 1 km in diameter by 2008—a goal met by 2011. In 2005, Congress passed the George E. Brown, Jr. Near-Earth Object Survey Act, requiring NASA to extend this mandate to include NEOs as small as 140 meters in diameter. By 2020, less than half of these smaller objects had been detected, prompting the development of new technologies to speed up detection.

To address the detection gap, NASA authorized the construction of the NEO Surveyor, a space-based infrared telescope designed to track potentially hazardous asteroids. Expected to launch in 2026, the NEO Surveyor is projected to dramatically shorten the time required to meet NASA's detection goals from 30 years to just 10 years, enhancing the world's planetary defense capabilities.

Since the 1990s, NASA has led various NEO detection programs using Earth-based observatories to identify and track asteroids. However, the limitations of terrestrial observation—such as difficulty in detecting dark asteroids near the Sun—prompted the need for space-based telescopes. These efforts culminated in the detection of thousands of NEOs, particularly those larger than 1 km in diameter, a size that poses a threat of global devastation upon impact.

In addition to ground-based efforts, NASA's infrared space telescope mission WISE, repurposed as NEOWISE in 2013, continues to scan the sky for potentially hazardous asteroids. The NEOWISE mission has been extended through 2023, providing critical data on NEOs that orbit close to Earth.

NASA's commitment to planetary defense has been integrated into several robotic missions. The NEAR Shoemaker spacecraft, launched in 1999, became the first spacecraft to orbit and land on an asteroid—433 Eros—in 2000. This mission greatly expanded our knowledge of asteroid composition and structure, offering insights into potential asteroid deflection strategies.

The OSIRIS-REx mission to asteroid Bennu, which launched in 2016, further enhanced our understanding of NEOs. Bennu, a near-Earth asteroid with potential future close encounters with Earth, was studied for its composition, orbit, and physical properties. The precision of the OSIRIS-REx measurements will enable scientists to better predict Bennu's orbit and its future interactions with Earth.

Another breakthrough occurred with the Double Asteroid Redirection Test (DART), a joint mission by NASA and the Johns Hopkins Applied Physics Laboratory (JHAPL). Launched in November 2021, DART was designed to test whether a spacecraft could successfully alter the orbit of an asteroid through kinetic impact. On September 26, 2022, DART collided with Dimorphos, a moonlet orbiting the asteroid Didymos. The mission exceeded expectations, shortening Dimorphos' orbital period by 32 minutes, a major success in demonstrating the potential for asteroid deflection.

In June 2022, NASA formally launched an independent study into Unidentified Aerial Phenomena (UAP), a topic previously investigated by the Pentagon and intelligence agencies. Spearheaded by NASA's Science Mission Directorate, the study is an attempt to apply rigorous scientific analysis to the observation and study of UAPs. Thomas Zurbuchen, head of the directorate, acknowledged that while the subject may be controversial, NASA's objective is to bring a data-driven approach to the study of these phenomena. This high-risk, high-reward initiative aims to provide new insights into unexplained aerial sightings through the lens of scientific inquiry.

In doing so, NASA is leveraging its expertise in satellite observation and atmospheric science to better understand UAPs. This research aligns with the agency's broader goals of improving the detection of natural and artificial phenomena in Earth's atmosphere and space.

NASA has a long history of collaboration with both domestic and international agencies to advance its goals in space exploration, research, and technology development. These partnerships have been instrumental in NASA's ability to conduct complex missions, share scientific data, and develop cutting-edge technologies for space and Earth sciences.

In the wake of the Apollo 1 accident in 1967, where three astronauts tragically lost their lives, Congress mandated the creation of the Aerospace Safety Advisory Panel (ASAP) to advise NASA's Administrator on safety-related issues in space programs. ASAP became a crucial voice in identifying hazards and improving safety protocols for NASA's air and space endeavors. Following the Space Shuttle Columbia disaster in 2003, Congress required that ASAP submit annual reports directly to the NASA Administrator and Congress, further emphasizing its critical role in space safety oversight.

By 1971, NASA had also formed two additional advisory bodies: the Space Program Advisory Council and the Research and Technology Advisory Council. These councils provided guidance on broader issues related to space missions and technological innovation. In 1977, these two councils were merged to form the NASA Advisory Council (NAC), consolidating NASA's advisory resources under one body. The NAC continues to play an essential role in advising the agency on various aspects of space exploration, technology, and safety, with the NASA Authorization Act of 2014 reaffirming the importance of the ASAP's oversight.

NASA and the National Oceanic and Atmospheric Administration (NOAA) have collaborated for decades in developing, launching, and operating weather satellites that provide vital information for global weather forecasting. NASA typically takes responsibility for these satellites' design, construction, and launch, while NOAA operates them and delivers weather products to users worldwide.

NOAA's Polar Operational Environmental Satellites (POES) have provided critical data for weather imaging from low Earth orbit, while the Geostationary Operational Environmental Satellites (GOES) deliver near-real-time coverage of the Western Hemisphere, offering unparalleled insight into developing weather phenomena. This longstanding partnership has been fundamental to improving the accuracy and timeliness of global weather forecasting.

The relationship between NASA and the U.S. military, particularly with the United States Space Force (USSF) and its predecessor in the Air Force, has been a cornerstone of U.S. space activities. NASA relies on the Space Force for essential support during launches from Kennedy Space Center, Cape Canaveral Space Force Station, and

Vandenberg Space Force Base. The Space Force provides critical services, such as range safety and rescue operations, ensuring that NASA's missions proceed smoothly and safely.

NASA and the Space Force have also partnered in areas such as planetary defense, including collaborations on asteroid impact mitigation strategies. Additionally, Space Force members are eligible to become NASA astronauts. For example, Colonel Michael S. Hopkins, the commander of SpaceX Crew-1, was commissioned into the Space Force while aboard the International Space Station in December 2020. This collaboration was formalized in a memorandum of understanding signed in 2020, cementing the cooperative relationship between the two organizations.

NASA's partnership with the U.S. Geological Survey (USGS) dates back to the 1970s with the launch of the Landsat program, the longest-running enterprise for acquiring satellite imagery of Earth. The first satellite in the series, Landsat 1, was launched on July 23, 1972, marking the beginning of a mission to collect detailed images of Earth's surface. These images have been vital in agriculture, cartography, forestry, and environmental monitoring.

The latest satellite, Landsat 9, launched on September 27, 2021, continues this legacy, with NASA responsible for building and launching the spacecraft, while the USGS manages its operations. The Landsat imagery archive, accessible through the USGS's EarthExplorer platform, has become an indispensable resource for global change research.

NASA and the European Space Agency (ESA) have enjoyed a close and productive partnership spanning decades. From the collaborative Spacelab missions aboard the Space Shuttle to joint ventures such as the Hubble Space Telescope and the James Webb Space Telescope, NASA and ESA have advanced the frontiers of space science together.

One of the most notable collaborations in recent years is the Artemis Gateway, a critical part of NASA's Artemis program aimed at returning humans to the Moon. ESA's contributions to the Gateway include modules for habitation and refueling, as well as enhancements to lunar communications. This partnership continues to play a key role in NASA's plans for sustainable lunar exploration and future missions to Mars.

NASA's collaboration with the Japan Aerospace Exploration Agency (JAXA) has been instrumental in advancing space exploration, particularly in Earth observation and the Artemis program. JAXA is a key contributor to the Lunar Gateway, providing vital systems for life support, thermal control, and other critical technologies.

The NASA-JAXA Global Precipitation Measurement (GPM) mission, launched in 2014, is another shining example of their successful collaboration. This mission has provided scientists and forecasters with highly accurate, near-real-time measurements of global precipitation, improving weather prediction and climate research.

NASA's relationship with Russia's space agency, Roscosmos, has been essential to the success of the International Space Station (ISS) since 1993. Despite geopolitical challenges, NASA and Roscosmos have cooperated in launching, maintaining, and operating the ISS, with both countries' astronauts working side by side aboard the station.

Even after the Space Shuttle retired in 2011, Russia's Soyuz spacecraft continued to serve as the only means of transporting crew to and from the ISS until the advent of NASA's Commercial Crew Program. In July 2022, NASA and Roscosmos signed an agreement to share ISS flight opportunities, enabling astronauts from both countries to continue flying on each other's spacecraft.

In 2014, NASA and the Indian Space Research Organisation (ISRO) signed a partnership to collaborate on a joint radar mission known as the NASA-ISRO Synthetic Aperture Radar (NISAR). Scheduled for launch in 2024, NISAR will provide critical data on global environmental changes, such as deforestation, ice-sheet collapse, and natural disasters. This collaboration highlights NASA's commitment to leveraging international partnerships to address global challenges.

The Artemis Accords, initiated by NASA and the U.S. State Department, represent a framework for peacefully exploring the Moon, Mars, asteroids, and other celestial bodies. These bilateral agreements define the principles for responsible behavior in space exploration and resource use. As of September 2022, 21 nations have signed the accords, demonstrating broad international support for NASA's vision of sustainable space exploration.

China National Space Administration (CNSA)

Due to the Wolf Amendment passed by Congress in 2011, NASA is prohibited from engaging in direct bilateral cooperation with China or China-affiliated organizations, such as the China National Space Administration (CNSA), without explicit approval from Congress. This law has been renewed annually and continues to limit NASA's ability to collaborate with China, despite China's growing presence in space exploration.

Space domain awareness (SDA) refers to the comprehensive study and monitoring of artificial objects in orbit around the Earth. This critical discipline encompasses detecting, tracking, cataloging, and identifying active and inactive satellites, spent rocket stages, and fragmentation debris. With thousands of objects in Earth's orbit, ensuring a clear understanding of their positions and trajectories is essential for global security, scientific research, and protecting valuable assets in space.

The primary objectives of space domain awareness are both practical and strategic. SDA helps predict when and where space objects, such as decommissioned satellites or spent rocket bodies, will re-enter the Earth's atmosphere. This information is vital for mitigating the risk of space debris causing harm upon re-entry. Additionally, SDA plays a crucial role in preventing false alarms in missile attack warning systems by distinguishing between decaying space objects and potential missile threats. Through detailed tracking and analysis, space domain awareness allows for the charting of current positions and future orbital paths of objects, ensuring that satellites, space stations, and other spacecraft can operate safely without collisions or interference.

Another key aspect of SDA is the ability to detect new man-made objects in space and continuously update a comprehensive catalog of space debris and satellites. This helps determine the country of origin for re-entering objects, allowing for international accountability. It also provides critical data that could inform the development of anti-satellite weapons systems, although this remains a controversial aspect of SDA given the potential for weaponization of space.

Space domain awareness is supported by various sophisticated systems, many of which are ground-based, while others involve space-borne assets. These systems span multiple nations, reflecting the global nature of space activities.

One of the primary systems is the United States Space Surveillance Network (SSN), which utilizes a range of detectors and sensors to track space objects. The SSN has incorporated the Space Fence, a high-precision radar system designed to replace the older Air Force Space Surveillance System. It also includes the Space Surveillance Telescope, a cutting-edge facility capable of tracking deep-space objects with remarkable accuracy.

Russia's Centre for Outer Space Monitoring employs a combination of advanced systems, such as Okno and Krona, along with the Sazhen-S radar, to monitor and analyze the orbits of spacecraft and debris. These systems allow Russia to maintain situational awareness of space objects in Earth's orbit.

In Europe, the French GRAVES (Grand Réseau Adapté à la Veille Spatiale) system, operated by the French Air Force, plays a significant role in space surveillance. This bi-static radar-based system tracks space objects over French airspace, contributing to Europe's broader space domain awareness efforts. The European Space Situational Awareness Programme further extends Europe's capabilities, with multiple assets providing comprehensive surveillance through its Space Surveillance and Tracking Segment.

Australia has also entered the space domain awareness field with its Silentium Defence Oculus radar-based observatory in Swan Reach, South Australia. This passive radar system enhances Australia's ability to monitor objects in space without active transmissions that could interfere with space operations.

One of the more innovative contributions to space domain awareness is the DEEP-Sight prototype, developed by Deepinder Uppal for the U.S. Defense Innovation Unit. This advanced system is designed to provide broad-spectrum SDA data to the United States Space Force, Navy, and public satellite research organizations, offering a more versatile and integrated approach to space monitoring.

Lastly, the National Reconnaissance Office (NRO) operates Skybarker satellites, a network of satellites that assist with space surveillance and reconnaissance, contributing to the overall U.S. capability in space domain awareness.

As the number of satellites and space objects continues to grow, space domain awareness becomes increasingly critical for the safe and secure use of space. Collaboration between nations and agencies worldwide highlights the importance of monitoring and managing the space environment. SDA systems ensure that the growing number of satellites and other objects can coexist without leading to catastrophic collisions while also providing the strategic advantage of knowing precisely what is happening in Earth's orbit at any given moment.

Chapter 7 The National Reconnaissance Office (NRO)

The National Reconnaissance Office (NRO), a key agency within the Department of Defense and the U.S. Intelligence Community, is tasked with designing, constructing, launching, and maintaining intelligence-gathering satellites. As part of its mission, the NRO provides critical support to national security through its advanced satellite technologies, offering invaluable intelligence, surveillance, and reconnaissance capabilities. The U.S. Space Force (USSF), playing an essential role in executing the NRO's space launches, comprises a significant portion—about 40%—of the NRO's personnel. The close collaboration between the two entities strengthens the overall effectiveness of space-based intelligence operations.

In recent years, proposals have emerged, including suggestions from the Air Force Association and retired Air Force Lieutenant General David Deptula, to further consolidate national security space operations by merging the NRO into the Space Force. This proposal envisions the NRO becoming a part of a Space Force Intelligence, Reconnaissance, and Surveillance Command, which would centralize the U.S. national security space apparatus within the USSF. Such a move would create a unified command structure, ensuring seamless coordination of space-based intelligence assets.

One of the most prominent responsibilities of the USSF in relation to the NRO is its management of the National Security Space Launch (NSSL) program. This program, overseen by the USSF's Space Systems Command (SSC) in partnership with the NRO, involves the use of both government and contracted spacecraft to launch sensitive and high-priority payloads into orbit. The NSSL serves not only the needs of the Space Force but also those of the NRO, ensuring that critical defense and intelligence satellites are placed into orbit safely and efficiently. NRO Director Christopher Scolese has emphasized the NRO's indispensable role in maintaining American space dominance, noting that the agency delivers "unrivaled situational awareness and intelligence" through the best imagery and signals intelligence on Earth.

Further strengthening this partnership, Lt. Gen. Michael Guetlein, a former deputy director of the NRO, assumed command of the USSF's Space Systems Command in August 2021. His leadership further reinforces the deep connection between the NRO and the Space Force, highlighting the ongoing integration of intelligence and defense space operations as they work together to protect U.S. interests in space and maintain a technological edge in this critical domain.

The National Reconnaissance Office (NRO), an integral part of the United States Intelligence Community, operates under the United States Department of Defense. Since its inception, the NRO has been responsible for the design, construction, launch, and operation of the U.S. government's reconnaissance satellites. These satellites provide critical intelligence across various domains. Notably, the NRO supplies signals intelligence (SIGINT) to the National Security Agency (NSA), imagery intelligence (IMINT) to the National Geospatial-Intelligence Agency (NGA), and measurement and signature intelligence (MASINT) to the Defense Intelligence Agency (DIA).

In 2023, the NRO announced its ambitious plan to significantly expand its satellite operations. Over the following decade, the agency intends to quadruple the number of satellites it manages, with the goal of delivering ten times the current volume of signals and imagery intelligence to its users. This growth underscores the NRO's pivotal role in national security as technological advancements in space-based reconnaissance continue to evolve.

As one of the "big five" U.S. intelligence agencies, alongside the CIA, NSA, DIA, and NGA, the NRO's influence and reach are significant. Headquartered in Chantilly, Virginia, near Washington Dulles International Airport, the NRO's operations are overseen by both the Director of National Intelligence and the Secretary of Defense. The agency's workforce consists of approximately 3,000 personnel, comprising a blend of NRO cadre and representatives from various branches of the U.S. military, the CIA, NSA, NGA, and U.S. Space Force. Despite its relatively small federal workforce, the NRO's operations rely heavily on defense contractors, who have been integral to its mission since the organization's early years.

The NRO was founded in 1960 following a recommendation to President Dwight D. Eisenhower amid concerns about the management of the U.S. Air Force's reconnaissance satellite programs. This new agency was established to unify the reconnaissance efforts of the Air Force, CIA, and other emerging intelligence organizations. The NRO's existence remained a closely guarded secret for decades, only being officially acknowledged in 1992 after years of speculation.

The NRO's early efforts were embodied in the Corona program, the first U.S. photo-reconnaissance satellite initiative. Launched in 1960, Corona was revolutionary for its time, using film-return capsules to capture high-resolution images of strategic areas, including Soviet missile sites. These capsules, once ejected from the satellite, were recovered mid-air by specially equipped aircraft. The Corona program continued until 1972, completing 145 successful missions. This program marked the first time the U.S. had gained consistent intelligence from space, laying the foundation for future satellite reconnaissance systems.

In addition to the Corona program, the NRO also conducted mapping missions under the Argon and Lanyard programs, though these were met with limited success. As technology progressed, the NRO's satellite missions became increasingly classified, and many of the details surrounding their operations remain undisclosed to this day.

Over the years, the NRO has faced challenges beyond its technical endeavors. In the mid-1990s, the agency became embroiled in a funding controversy when it was revealed that the NRO had accumulated a surplus of unspent funds, estimated to be as high as $1.7 billion. This financial reserve had been built without the knowledge of the CIA, the Pentagon, or Congress, prompting inquiries into the agency's accounting practices. Investigations revealed that the NRO had used part of these funds to construct its headquarters in Chantilly, Virginia. As a result of this controversy, the NRO merged its three separate accounting systems to improve financial oversight.

One of the NRO's most ambitious but ultimately unsuccessful projects was the Future Imagery Architecture (FIA) program. Launched in 1999 in partnership with Boeing, the FIA was intended to develop a new generation of imaging satellites. However, the project faced numerous technical setbacks and cost overruns, eventually becoming one of the most expensive failures in the history of U.S. intelligence. By 2005, after billions of dollars had been spent, the FIA program was canceled, a stark reminder of the complexity and risk involved in cutting-edge satellite technology development.

In the early 2000s, the National Reconnaissance Office (NRO) found itself at the center of a significant security breach. On August 23, 2001, Brian Patrick Regan, a civilian contractor employed by TRW at the NRO, was arrested at Dulles International Airport as he prepared to board a flight to Zurich. Regan was carrying coded information on Iraqi and Chinese missile sites, along with the addresses of their embassies in Switzerland and Austria. He had attempted to sell these sensitive intelligence secrets to Iraq and China, an act that ultimately led to his life sentence without parole.

In January 2008, the NRO faced another crisis when it was revealed that one of its reconnaissance satellites, believed to be USA-193, was malfunctioning and would re-enter Earth's atmosphere uncontrollably. The satellite, built by Lockheed Martin and launched in December 2006, had failed shortly after reaching orbit. To prevent the potential dangers of an uncontrolled re-entry, the Pentagon took the unprecedented step of shooting down the satellite. On February 21, 2008, a Navy missile successfully intercepted the satellite, breaking it into multiple fragments.

As satellite technology continued to advance, the NRO declassified the existence of its Synthetic Aperture Radar satellites in July 2008. This decision was made to facilitate discussions with other branches of the military, particularly the U.S. Air Force, regarding space-based radar systems. Around the same time, the NRO became a non-voting member of the Civil Applications Committee (CAC), an inter-agency group that coordinated the federal use of classified satellite imagery for civil applications.

Throughout the 2010s, the NRO continued to face challenges, both in terms of operational complexity and internal controversies. In 2012, a McClatchy investigation raised concerns about the ethical practices of the NRO's polygraph examiners, who were accused of improperly gathering personal information from Department of Defense personnel. In a more serious case, an inspector general's report in 2014 revealed that the NRO had failed to report admissions of child sexual abuse gathered during polygraph testing to law enforcement.

On the operational front, the NRO remained at the cutting edge of reconnaissance technology. In 2010, NRO Director General Bruce Carlson announced that the agency was entering its busiest launch period in 25 years, with a series of large and critical reconnaissance satellites planned for orbit. These launches were crucial for maintaining the United States' dominance in space-based intelligence gathering.

One of the most publicized incidents involving the NRO occurred in 2019 when then-President Donald Trump tweeted an image of the aftermath of a failed Iranian rocket launch. The image, almost certainly from the NRO's USA-224 satellite, revealed the extent to which the NRO's surveillance capabilities had advanced.

In the 2020s, the NRO continued its aggressive satellite launch schedule, deploying numerous classified payloads aboard various rockets, including SpaceX's Falcon 9, Northrop Grumman's Minotaur, and United Launch Alliance's Delta IV. These launches included high-profile missions such as NROL-108 in December 2020, NROL-82 in April 2021, and NROL-91 in September 2022, all of which contributed to expanding the United States' reconnaissance capabilities.

Looking ahead, the NRO's partnership with SpaceX marks a new era in satellite intelligence. In 2021, SpaceX secured a $1.8 billion contract from the NRO to build a network of spy satellites under its Starshield unit. These satellites are designed to track ground targets and provide continuous imagery to U.S. military and intelligence officials, significantly enhancing real-time surveillance capabilities across the globe.

The National Reconnaissance Office (NRO), an essential component of the Department of Defense, plays a pivotal role in U.S. national security through its management of reconnaissance satellites. The Director of the NRO, appointed by the President and confirmed by the Senate in accordance with Title 50 of the U.S. Code, oversees the agency's operations. Historically, this position was filled by the Under Secretary of the Air Force or the Assistant Secretary of the Air Force for Space. However, in 2005, the role became independent with the appointment of Donald Kerr as Director.

The NRO's organization reflects its complex responsibilities, with multiple directorates and offices managing various aspects of satellite intelligence gathering and operations:

Principal Deputy Director of the NRO (PDDNRO): This position supports the Director in managing all NRO activities and assumes decision-making responsibilities as delegated. In the Director's absence, the PDDNRO acts on their behalf.

Deputy Director of the NRO (DDNRO): Traditionally filled by a senior U.S. Air Force (USAF) general officer, the DDNRO represents military personnel within the NRO and assists in coordinating NRO operations with the Air Force.

Corporate Staff: This encompasses legal, human resources, procurement, public affairs, and other essential support functions necessary for the NRO's day-to-day operation.

Several directorates handle the technical, operational, and business aspects of NRO's mission:

Office of Space Launch (OSL): Responsible for the coordination and execution of satellite launches, including mission assurance, operations, and safety. OSL also liaises with industry, NASA, and the USAF.

Advanced Systems and Technology Directorate (AS&T): Focused on developing cutting-edge technology, AS&T drives innovation in satellite reconnaissance systems.

Business Plans and Operations (BPO): Manages the NRO's financial, budgetary, and legislative communications, ensuring the efficient use of resources.

Ground Enterprise Directorate (GED): Provides the infrastructure for collecting and distributing satellite intelligence to users globally.

Geospatial Intelligence Systems Acquisition Directorate (GEOINT): Acquires advanced imagery systems for geospatial intelligence, supplying critical data to the Intelligence Community and military.

Mission Operations Directorate (MOD): Oversees the operation and maintenance of NRO satellites and ground systems. MOD manages the 24-hour NRO Operations Center, coordinating with U.S. Strategic Command to ensure satellite safety and space situational awareness.

Signals Intelligence Systems Acquisition Directorate (SIGINT): Builds and deploys satellites that gather communication, electronic, and foreign signals intelligence.

Mission Integration Directorate (MID): Engages with NRO's mission partners to address operational intelligence needs and provides solutions to military commanders through the Tactical Defense Space Reconnaissance (TacDSR) Program.

The NRO operates as a hybrid organization of approximately 3,000 personnel, which includes members of the armed forces, CIA, and Department of Defense (DoD) civilians. This workforce is expected to grow modestly, with an increase of 100 personnel projected between 2010 and 2012. The bulk of the NRO's workforce is comprised of private contractors, with $7 billion of its $8 billion budget allocated to corporate partnerships. This reliance on external contractors highlights the NRO's operational complexity and the importance of industry collaboration in advancing its mission.

The NRO's funding comes from both the U.S. intelligence budget and the military budget. Its budget has grown significantly over the decades, reflecting the increasing importance of space-based reconnaissance. In 1971, the NRO's annual budget was estimated at $1 billion ($7.5 billion in 2024 dollars). By 1994, this figure had risen to $6 billion ($12.3 billion in 2024 dollars). The NRO's budget for 2010 was projected to reach $15 billion ($21 billion in 2024 dollars), representing 19% of the overall U.S. intelligence budget of $80 billion for that fiscal year. In 2012, the NRO increased its allocation for science and technology, raising it to nearly 6% of its budget, after years of decline to just 3%.

In mid-2019, just before the formal establishment of the United States Space Command (USSPACECOM), Air Force General John W. Raymond, set to lead the new command, underscored the close collaboration between the National Reconnaissance Office (NRO) and USSPACECOM. Raymond emphasized that the NRO would respond directly to the commander's directives, particularly in safeguarding and defending the United States' space-based capabilities. He further noted that the NRO and USSPACECOM shared not only a common vision but also coordinated operations, training, and staffing at the National Space Defense Center, symbolizing the integration of military and intelligence efforts in space.

The relationship between the NRO and U.S. military space operations deepened with the establishment of the United States Space Force (USSF) in December 2019, led by General Raymond as Chief of Space Operations. The NRO maintained a strong partnership with the Space Force's Space and Missile Systems Center (SMC) through the National Security Space Launch (NSSL) program. This program ensured that essential government payloads were successfully launched into orbit, supporting both the NRO and the military. NRO Director Christopher Scolese highlighted the agency's vital role in ensuring U.S. space dominance, noting that the NRO provided unparalleled situational awareness and intelligence, along with the most advanced imagery and signals data available.

In August 2021, Scolese, Raymond, and USSPACECOM Commander General James Dickinson formalized their cooperation through the Protect and Defend Strategic Framework. This agreement outlined the coordination between the Department of Defense (DoD) and the intelligence community in space security, encompassing everything from acquisition to operational activities. The collaboration reflects the growing importance of integrating intelligence and military space operations to safeguard U.S. interests in space.

The NRO is renowned for operating some of the most advanced space technologies, often surpassing civilian equivalents. As early as the 1980s, NRO satellites and software were capable of determining the exact dimensions of a tank's gun barrel, exemplifying the precision and sophistication of its surveillance systems. In 2012, the NRO donated two space telescopes to NASA, both more advanced than the Hubble Space Telescope, highlighting the cutting-edge capabilities at the NRO's disposal.

In terms of technological innovation, the NRO is also developing an artificial intelligence (AI) system known as Sentient. This automated intelligence analysis system is designed to process vast amounts of data at machine speed, enabling rapid and more comprehensive analysis than human operators could achieve. Sentient aims to enhance intelligence collection by using automated inferencing to predict future events and direct NRO satellites to areas of interest. The system represents a leap forward in the use of AI for intelligence gathering, with the potential to revolutionize space-based reconnaissance.

The NRO manages four primary satellite constellations, each serving distinct intelligence functions:

Signals Intelligence (SIGINT) Constellation: Gathers communication and electronic intelligence from various global sources.

Geospatial Intelligence (GEOINT) Constellation: Provides high-resolution imagery for geospatial intelligence.

Communications Relay Constellation: Supports secure communication channels for intelligence and military operations.

Reconnaissance Constellation: Monitors global activities, ensuring continuous surveillance of areas of interest.

Among the NRO's most notable spacecraft are the Keyhole series of imagery intelligence satellites. These include the KH-1 through KH-9 programs, such as the Corona, Argon, and Hexagon systems, which were pivotal during the Cold War. The KH-11 series, known for its advanced electro-optical capabilities, has provided some of the clearest images from space since its deployment in 1976.

Other satellites include the Lacrosse/Onyx radar imaging satellites and the Poppy electronic intelligence (ELINT) program, which began in the 1960s. The NRO's Mentor and Orion satellites continue to gather high-altitude signals intelligence. The agency's classified NROL missions, numbering over 66 to date, represent a range of secret payloads launched into orbit.

The NRO's mission ground stations include facilities in the United States and abroad.

NRO Headquarters in Chantilly, Virginia, where the National Reconnaissance Operations Center (NROC) is housed, overseeing satellite operations and space situational awareness.

Aerospace Data Facility-Colorado (ADF-C) at Buckley Space Force Base, Colorado, which is a key ground station for NRO satellite data processing.

Aerospace Data Facility-East (ADF-E) at Fort Belvoir, Virginia, and Aerospace Data Facility-Southwest (ADF-SW) at White Sands, New Mexico.

International locations include RAF Menwith Hill in the UK and Pine Gap in Australia, both essential for global intelligence sharing and satellite communication.

These facilities provide vital support for the NRO's global surveillance efforts, ensuring that critical intelligence is processed and distributed in real-time to U.S. intelligence and military agencies. The NRO also maintains launch offices at Cape Canaveral Space Force Station in Florida and Vandenberg Space Force Base in California, facilitating the deployment of new satellites into orbit.

The NRO's coordination with USSPACECOM and the USSF, coupled with its advanced technologies, ensures that the United States maintains a leading role in space intelligence. With innovations like the Sentient AI system and an extensive array of reconnaissance satellites, the NRO continues to be a critical asset for both military operations and national security.

Chapter 8 - The US Department of Defense (DOD)

The United States Department of Defense (DoD) serves as the executive branch of the federal government responsible for overseeing all agencies and functions related to national security and the Armed Forces. As of November 2022, the DoD stands as the second-largest employer in the world, with over 2.91 million employees, including more than 1.4 million active-duty military personnel and hundreds of thousands of National Guard, reservists, and civilians. The department's mission is to ensure the military forces are equipped to deter war and safeguard national security. Headquartered at the Pentagon in Arlington, Virginia, just outside Washington, D.C., the DoD plays a critical role in the U.S. defense structure.

At the helm of the Department of Defense is the Secretary of Defense, a cabinet-level official reporting directly to the President of the United States, who also serves as Commander-in-Chief of the U.S. Armed Forces. Beneath the Secretary, the Department is divided into three primary branches: the Department of the Army, the Department of the Navy, and the Department of the Air Force. The department also oversees key national intelligence services such as the Defense Intelligence Agency (DIA), National Security Agency (NSA), National Geospatial-Intelligence Agency (NGA), and National Reconnaissance Office (NRO), along with other vital agencies, including the Defense Advanced Research Projects Agency (DARPA), the Missile Defense Agency (MDA), and the Defense Health Agency (DHA). Military operations are carried out by eleven unified combatant commands that manage operations worldwide.

The origins of the U.S. military can be traced back to the Revolutionary War. In response to escalating tensions with the British government, the First Continental Congress recommended that the Thirteen Colonies prepare for defensive military action in 1774. With the outbreak of war in June 1775, the Second Continental Congress established the Continental Army to form a unified national force capable of fighting beyond colonial boundaries. This event, marked by the Army's founding on June 14, 1775, is now celebrated annually as Flag Day. That same year, Congress chartered the Continental Navy on October 13 and the Continental Marines on November 10, laying the groundwork for a national defense structure.

Following the formation of the United States under the Constitution in 1789, the newly seated Congress was initially slow to organize military forces. It was only after repeated requests from President George Washington that Congress passed legislation on September 29, 1789, establishing the War Department, responsible for the Army and, until 1798, naval affairs. By 1798, a separate Navy Department was created to manage naval operations. Both departments operated independently, with their respective Secretaries reporting directly to the President.

In the aftermath of World War II, President Harry Truman sought to reorganize the U.S. military under a unified structure, citing inefficiencies and conflicts between military departments. This led to the passage of the National Security Act of 1947, a landmark piece of legislation that established the National Military Establishment (NME), which placed all branches of the military under the control of a single Secretary of Defense. The act also created the Central Intelligence Agency (CIA), the National Security Council, and the Joint Chiefs of Staff, marking the beginning of a more integrated national defense system.

The National Military Establishment began operations on September 18, 1947, with James V. Forrestal confirmed as the first Secretary of Defense. In 1949, the establishment was renamed the Department of Defense, and the separate military departments—Army, Navy, and Air Force—were brought under its unified command. This name change, partially driven by concerns over the abbreviation "NME" being pronounced as "enemy," solidified the Department of Defense's role as the central authority for U.S. military operations.

The Department of Defense continued to evolve through the 20th century, particularly with the Defense Reorganization Act of 1958. This act streamlined authority within the department, granting the Secretary of

Defense greater oversight over military departments while maintaining their responsibilities to organize, train, and equip forces. The legislation also established the chain of command for military operations, running from the President to the Secretary of Defense through the Unified Combatant Commanders. Additionally, the act centralized research efforts under the Advanced Research Projects Agency (ARPA), later known as DARPA, underscoring the Department's commitment to maintaining technological superiority.

The reorganization efforts, spearheaded by President Dwight D. Eisenhower's administration, marked a significant transformation in how the U.S. military functioned, ensuring a more cohesive and efficient defense system capable of responding to the complex global challenges of the Cold War era. These reforms would set the stage for the military's role in supporting America's strategic objectives during the Space Race and beyond.

The organizational structure of the United States Department of Defense (DoD) is designed to ensure efficient management and coordination of the nation's defense operations. At the head of the DoD is the Secretary of Defense, appointed by the President with the advice and consent of the Senate. As the chief advisor to the President on all defense-related matters, the Secretary of Defense holds authority, direction, and control over the entire department, as outlined by federal law (10 U.S.C. § 113). The Secretary's statutory powers stem from the constitutional authority of both Congress and the President, who hold ultimate responsibility for military matters. Given the complexity and scope of military affairs, the Secretary of Defense and subordinate officials exercise military authority on behalf of the President and Congress.

The Department of Defense is composed of several key organizational units, each serving a specific function within the defense apparatus. These include the Office of the Secretary of Defense (OSD), the Joint Chiefs of Staff (JCS) and their supporting Joint Staff (JS), the Office of the Inspector General (DODIG), the Combatant Commands, the Military Departments (Department of the Army, Department of the Navy, and Department of the Air Force), as well as various Defense Agencies and Field Activities. Additional entities, such as the National Guard Bureau and other offices and commands, are also part of the DoD framework as established by law, presidential orders, or directives from the Secretary of Defense.

The relationships and functions within the DoD are further defined in Department of Defense Directive 5100.01, which serves as the foundational document outlining the department's organizational responsibilities. The most recent revision of this directive, signed by Secretary of Defense Robert Gates in December 2010, represented the first significant update since 1987, reflecting the evolving needs of the U.S. defense structure.

The Office of the Secretary of Defense (OSD) is the primary policymaking and management body within the Department of Defense. Staffed predominantly by civilian personnel, the OSD oversees a wide range of functions including policy development, resource management, planning, program evaluation, and coordination with other U.S. federal agencies, foreign governments, and international organizations. The OSD also manages the Defense Agencies, Department of Defense Field Activities, and specialized cross-functional teams, providing oversight and guidance across the broader defense infrastructure.

Several defense agencies fall under the purview of the OSD, each fulfilling a specific role within the national defense apparatus. These agencies include the Defense Advanced Research Projects Agency (DARPA), which focuses on breakthrough military technologies, the Defense Logistics Agency (DLA), responsible for supply chain management and logistics, and the Defense Health Agency (DHA), which oversees medical services for military personnel. Other key agencies include the Defense Information Systems Agency (DISA), which handles global communication and information systems, and the Missile Defense Agency (MDA), charged with developing and deploying missile defense systems.

In addition to its defense responsibilities, the Department of Defense also plays a critical role within the United States Intelligence Community. Several defense agencies operate as national-level intelligence services under the Department's jurisdiction, while simultaneously coordinating with the Office of the Director of National

Intelligence. These agencies, such as the National Security Agency (NSA), the National Geospatial-Intelligence Agency (NGA), and the National Reconnaissance Office (NRO), serve both national policymakers and military commanders by providing intelligence critical to defense and national security operations.

The DoD is also responsible for managing intelligence assets in disciplines such as signals intelligence, geospatial intelligence, and measurement and signature intelligence. It oversees the development, launch, and operation of intelligence satellites, ensuring the U.S. retains strategic and tactical advantages in intelligence gathering. Additionally, the Department maintains its own human intelligence services, which work in conjunction with the Central Intelligence Agency (CIA) to support both military and civilian intelligence priorities. These agencies are overseen by the Under Secretary of Defense for Intelligence and Security, ensuring a unified approach to intelligence and security matters across the defense and intelligence communities.

The Joint Chiefs of Staff (JCS) is a body of the most senior military leaders within the Department of Defense, responsible for advising the President, the Secretary of Defense, the National Security Council, and the Homeland Security Council on military matters. The composition of the Joint Chiefs of Staff is established by law and includes the Chairman of the Joint Chiefs of Staff (CJCS), Vice Chairman of the Joint Chiefs of Staff (VCJCS), Senior Enlisted Advisor to the Chairman (SEAC), and the chiefs of the respective military services: the Army, Marine Corps, Navy, Air Force, Space Force, and the National Guard Bureau. All of these leaders are appointed by the President and confirmed by the U.S. Senate.

Each military service chief, while serving on the Joint Chiefs of Staff, retains responsibilities within their respective military departments. For example, the Chief of Staff of the Army works directly with the Secretary of the Army, the Chief of Naval Operations with the Secretary of the Navy, and so on. Despite their roles in advising on broader military issues, the service chiefs no longer hold operational command authority as individuals or as a group, following the enactment of the Goldwater–Nichols Act in 1986.

The Goldwater–Nichols Act fundamentally restructured the chain of command and the advisory responsibilities of the Joint Chiefs of Staff. This legislation designated the Chairman of the Joint Chiefs of Staff as the principal military advisor to the President, the National Security Council, the Homeland Security Council, and the Secretary of Defense. The other members of the Joint Chiefs can only provide their advice through the Chairman, ensuring a unified and coherent military perspective is presented to civilian leadership. Additionally, the Act established the position of Vice Chairman of the Joint Chiefs of Staff, who assists the Chairman in carrying out these advisory responsibilities.

This act further clarified operational control over military forces, establishing that the chain of command flows directly from the President to the Secretary of Defense and from there to the Combatant Commanders who lead the military's operational forces. This streamlined the decision-making process and reinforced civilian oversight of the military.

To assist the Chairman and Vice Chairman in their duties, the Joint Staff (JS) operates out of the Pentagon and is composed of personnel from all branches of the armed forces. The Joint Staff is led by the Director of the Joint Staff (DJS), a position typically held by a lieutenant general or vice admiral. The Joint Staff plays a crucial role in supporting the strategic planning, policy formulation, and operational coordination needed to fulfill the responsibilities of the Joint Chiefs.

The Department of Defense is divided into three major military departments:

Department of the Army, which oversees the United States Army.

Department of the Navy, responsible for the United States Navy and United States Marine Corps.

Department of the Air Force, which governs both the United States Air Force and the United States Space Force.

Each military department is led by a civilian Secretary—Secretary of the Army, Secretary of the Navy, and Secretary of the Air Force—who are appointed by the President and confirmed by the Senate. These secretaries have broad legal authority, under Title 10 of the United States Code, to manage the day-to-day affairs of their respective departments. However, by law, they are subordinate to the Secretary of Defense and, through delegation, to the Deputy Secretary of Defense.

The service chiefs (e.g., Chief of Staff of the Army, Commandant of the Marine Corps, Chief of Naval Operations, and Chief of Space Operations) manage their branches' training, equipment, and administrative functions but do not have direct operational command authority over the troops. This distinction was formalized by the Defense Reorganization Act of 1958, which stripped operational control from the military departments, placing it instead with the unified Combatant Commands. As a result, the military departments focus on preparing and equipping the forces, while operational command lies with the Combatant Commanders during missions and engagements.

This carefully structured division of authority between operational and administrative roles ensures that while military leadership plays a crucial advisory role in national security and defense planning, the execution of military operations remains under unified command, ultimately ensuring efficiency and clarity in the chain of command.

The Unified Combatant Commands (UCCs) of the United States Department of Defense (DoD) represent the operational command structures responsible for directing military forces in various missions worldwide. A unified combatant command is a military command that includes personnel and equipment from at least two different military branches. These commands are tasked with broad and ongoing missions, often involving complex, cross-service coordination. While the military departments (Army, Navy, Air Force) focus on equipping and training their respective forces, the Unified Combatant Commands are responsible for the operational command of these forces during missions.

In military operations, the chain of command flows directly from the President to the Secretary of Defense, and then to the Combatant Commanders who lead the Unified Commands. Almost all U.S. military operational forces fall under the authority of one of these commands, which are governed by a Unified Command Plan. This plan, regularly updated by the Department of Defense, outlines the mission, geographical or functional responsibilities, and force structure of each command. The plan ensures that the right forces are assigned to each command and that there is clarity in terms of their missions and areas of responsibility.

As of 2019, the United States has eleven Unified Combatant Commands, divided into two categories: geographical commands, which are responsible for specific regions of the world, and functional commands, which focus on specialized capabilities across multiple regions.

U.S. Northern Command (USNORTHCOM): Responsible for defense and military operations within North America, including homeland defense and disaster response within the United States and its territories.

U.S. Southern Command (USSOUTHCOM): Focuses on operations in Latin America and the Caribbean, dealing with issues such as regional security, counter-narcotics, and disaster relief.

U.S. Central Command (USCENTCOM): Oversees military operations in the Middle East and parts of North Africa and Central Asia, with a primary focus on counter-terrorism and regional stability.

U.S. European Command (USEUCOM): Manages military operations across Europe, working closely with NATO allies to ensure security and cooperation in the region.

U.S. Indo-Pacific Command (USINDOPACOM): The largest combatant command by geographic scope, USINDOPACOM covers operations in the Indo-Pacific region, emphasizing security partnerships and freedom of navigation in areas such as the South China Sea.

U.S. Africa Command (USAFRICOM): Responsible for operations across the African continent, with a focus on counter-terrorism, peacekeeping, and building regional stability through partnerships.

U.S. Strategic Command (USSTRATCOM): Oversees the United States' nuclear forces, missile defense, and global strike capabilities, ensuring strategic deterrence and protection against major threats.

U.S. Special Operations Command (USSOCOM): Manages all special operations forces across the military, coordinating missions such as counter-terrorism, direct action, and unconventional warfare.

U.S. Transportation Command (USTRANSCOM): Provides transportation and logistics support for U.S. military operations globally, ensuring rapid mobility of forces and equipment during both peacetime and conflict.

U.S. Cyber Command (USCYBERCOM): Responsible for defending the United States against cyber threats, conducting cyber operations, and ensuring the security of U.S. military networks and systems.

U.S. Space Command (USSPACECOM): The newest combatant command, USSPACECOM oversees space operations, ensuring the security and defense of U.S. interests in space, including satellite systems and space-based communications.

Each of these combatant commands plays a crucial role in protecting U.S. interests and maintaining global security. Geographical commands focus on maintaining regional stability, fostering partnerships, and responding to crises within their respective areas of responsibility (AOR). For instance, USNORTHCOM is tasked with defending the U.S. homeland, while USEUCOM works with NATO allies to bolster security in Europe. Functional commands, on the other hand, provide specialized capabilities that span multiple regions, such as USTRANSCOM's global logistics network or USSOCOM's special operations expertise.

Together, these commands form the operational backbone of the U.S. military, ensuring that forces are well-equipped and effectively coordinated to meet the demands of both regional and global challenges.

Chapter 9 - US Army & Navy Space Commands

The Naval Space Command (NSC)

When the U.S. Space Force was officially established in 2019, one of its primary goals was to consolidate the existing military space forces from across the Army, Navy, and Air Force into a single, unified branch. This consolidation aimed to centralize the command and operations of all space-related military assets under the newly created Space Force, allowing for more cohesive and streamlined management of U.S. space capabilities.

The transition saw the Navy and Air Force willingly transfer their space forces to the Space Force, relinquishing control of their respective space-based operations in the interest of fostering greater coordination and efficiency. The Air Force, in particular, had long been the dominant player in military space operations through its Air Force Space Command, which became the foundation for the Space Force.

However, the transfer of space forces from the U.S. Army faced greater resistance. The Army, with its own well-established space operations supporting ground forces through satellite communications, missile defense, and intelligence gathering, was initially hesitant to cede control of these assets. Army leadership expressed concerns about maintaining the close integration of space operations with its terrestrial missions, fearing that the transfer of space functions might undermine the Army's ability to directly support its own operational needs.

This resistance highlighted the challenge of consolidating space forces under the Space Force while ensuring that the individual service branches could continue to receive the space-based support they required. Despite these concerns, discussions and negotiations continued as part of the broader effort to fully unify U.S. military space operations under the Space Force banner

The Naval Space Command (NSC) was a pivotal military command of the United States Navy, formed as a component of the U.S. Space Command. Established on October 1, 1985, and headquartered at Dahlgren, Virginia, the NSC was tasked with leveraging space capabilities to enhance naval operations. The command's primary responsibilities included operating reconnaissance and communication satellites, which supported naval forces, and advocating for the Navy's space interests within both the Navy and U.S. Space Command.

In July 2002, the Naval Space Command was merged into the Naval Network and Space Operations Command, which later became part of the broader Naval Network Warfare Command, marking a significant reorganization in the Navy's space and network operations.

The origins of Naval Space Command's mission trace back to the late 1950s, when the United States Naval Research Laboratory embarked on Project Vanguard. As part of this initiative, the Minitrack system was developed to detect and track satellites by using electronic signals. It was notably used to track Sputnik, the first artificial satellite launched by the Soviet Union in 1957, as well as subsequent satellites. This early satellite tracking effort provided one of the first ground-based systems to monitor objects in space. The Minitrack system evolved over time and, by 1961, became the Naval Space Surveillance System (NAVSPASUR). This system played a critical role in monitoring the orbits of satellites and was a key component of the Navy's space operations.

In 1993, the Naval Space Surveillance System was formally transferred to Naval Space Command. The system was operated through a network of transmitter and receiver sites across the United States. Three transmitter sites were located at Jordan Lake, Alabama, Lake Kickapoo, Texas, and Gila River, Arizona. Additionally, six receiver sites were located at Tattnall County, Georgia, Hawkinsville, Georgia, Silver Lake, Mississippi, Red River, Arkansas, Elephant Butte, New Mexico, and San Diego, California. This network provided comprehensive coverage for tracking satellite movements, allowing the Navy to maintain situational awareness of space activities.

In 1987, Naval Space Command expanded its mission by assuming responsibility for the Alternate Space Operations Center for U.S. Space Command. In this capacity, NSC provided backup space operations capabilities and continued its role in managing space surveillance networks. The command was instrumental in providing space intelligence support to naval forces, ensuring that the Navy had access to critical satellite data for strategic and operational planning.

Naval Space Command's space surveillance efforts were integral to the U.S. Space Command's broader mission of space situational awareness, and its operations were closely aligned with those of other military branches tasked with space-related responsibilities.

Although Naval Space Command was officially disestablished in 2002, its contributions to U.S. military space operations left a lasting impact. In 2004, many of NSC's functions, including the Naval Space Surveillance System, were transferred to the Air Force Space Command, specifically the 20th Space Control Squadron, which is now part of the United States Space Force. This transfer marked the formal integration of Navy space operations into the broader U.S. military space efforts led by the Air Force and, later, the Space Force.

The Naval Space Surveillance System, renamed the Air Force Space Surveillance System, continued to play a critical role in space monitoring and satellite tracking. Additionally, the responsibility for operating the Alternate Space Operations Center was also transferred, further consolidating space operations under the purview of Air Force Space Command.

Though no longer a standalone command, Naval Space Command's legacy remains through its pioneering efforts in space surveillance and its role in the development of modern military space operations.

US Army Space and Missile Defense Command

The United States Army Space and Missile Defense Command (USASMDC), currently serving as the Army Service Component Command (ASCC) for both the United States Strategic Command and the United States Space Command, has a long and intricate history rooted in the evolution of missile defense and space operations. Originally established in 1985 as the Army Strategic Defense Command, its initial focus was on ballistic missile defense. By 1992, the Army merged its Space Command with the Strategic Defense Command to form the Army Space and Strategic Defense Command, reflecting its growing commitment to space operations. In 1997, the organization was redesignated as the Army Space and Missile Defense Command, elevating it to the status of a Major Army Command.

The USASMDC plays a critical role in providing space, missile defense, and high-altitude capabilities to the Army. Its operational structure consists of two key components: the 100th Missile Defense Brigade, responsible for operating the Ground-Based Midcourse Defense system, and the 1st Space Brigade, tasked with providing trained space forces for global space support, enhancement, and control missions. The commander of the USASMDC also holds a pivotal role as the commander of the Joint Functional Component Command for Integrated Missile Defense under the U.S. Space Command.

In recent years, the Army's role in space operations has come under scrutiny, particularly following the establishment of the U.S. Space Force in 2019. While the Space Force absorbed the Navy's space forces and assumed responsibility for satellite communications and other space operations, the Army retained control of key elements such as the 100th Missile Defense Brigade and a scaled-down version of the 1st Space Brigade. However, the transfer of the Army's satellite communications mission to the Space Force in 2022 has sparked ongoing discussions about whether the remaining Army space assets should follow suit.

The origins of the Army's involvement in missile defense trace back to the late 1950s with the Nike Zeus program, an early effort to counter Soviet ballistic missiles. Although the program was not deployed, it laid the groundwork for subsequent initiatives such as the Nike-X and Sentinel programs. The Sentinel program, launched in 1967, represented an ambitious attempt to create a comprehensive missile defense network. However, amid escalating resource demands for the Vietnam War and public controversy, the Sentinel program was suspended in 1969 and replaced by the Safeguard program, which was more narrowly focused on defending key U.S. missile sites against limited attacks.

The Safeguard program's legacy is marked by the 1972 Anti-Ballistic Missile Treaty, which restricted the United States and the Soviet Union to two anti-ballistic missile sites, later reduced to one in 1974. The only Safeguard site to reach full operational capability was at what is now Cavalier Space Force Station, which remained functional until the program's cancellation in 1975. The site's Perimeter Acquisition Radar was repurposed for NORAD's early warning system, and it continues to operate under the Space Force today.

The launch of President Ronald Reagan's Strategic Defense Initiative (SDI) in 1983 reinvigorated the Army's missile defense activities. The Army Strategic Defense Command, established in 1985, played a leading role in ground-based interceptor programs, while the Air Force focused on space-based elements. Following the dissolution of the Strategic Defense Initiative in the post-Cold War era, the Army redirected its attention to theater missile defense, ultimately leading to the development of the Ground-Based Midcourse Defense system in the aftermath of the September 11 attacks.

In 1997, Army Space Command, which had continued as a sub-command within the broader Army Space and Strategic Defense Command, was restructured and renamed as Army Space and Missile Defense Command (SMDC). With this redesignation, the Army solidified its role in missile defense, as well as its broader space operations mission. The creation of the U.S. Army Futures Command in 2018 placed renewed emphasis on modernizing air and missile defense systems, particularly in areas such as hypersonics, short-range air defense (M-SHORAD), and integrated missile defense systems like the Integrated Air and Missile Defense Battle Command System (IBCS).

The establishment of the U.S. Space Force in 2019 marked a significant turning point for military space operations. Although the consolidation of space assets was central to the Space Force's mission, the Army initially resisted transferring its space units, which included the 1st Space Brigade, the 100th Missile Defense Brigade, and the Army Satellite Operations Brigade. Despite this resistance, the Army Satellite Operations Brigade was officially transferred to the Space Force on 15 August 2022, along with the Army's satellite communications mission, which was absorbed by Space Delta 8.

The debate surrounding the transfer of the Army's missile warning mission has been particularly contentious. While the Space Force assumed responsibility for global missile warning, the Army continued to operate regional missile warning through the Joint Tactical Ground Station (JTAGS) under the 1st Space Brigade. However, as of January 2023, the Army agreed to transfer the JTAGS mission to the Space Force's Space Delta 4, with the transition scheduled for completion by October 2023.

There have been continued calls for the complete transfer of the Army's space assets, including the 100th Missile Defense Brigade, to the Space Force. Proponents argue that the consolidation of missile defense and space operations would streamline command and control, while critics suggest that the Army's focus on integrating space capabilities with ground forces remains a vital component of land-based operations. In 2024, the Army outlined its vision for space operations, emphasizing the integration of space capabilities to support land operations, counter-satellite communications, and conduct navigation warfare.

As the debate continues, the Army remains committed to redefining its role in space, even as the Space Force expands its purview over U.S. military space operations. The evolution of the Army's involvement in space and missile defense underscores the complex and often contentious relationship between the services in the rapidly changing landscape of modern military space operations.

Chapter 10 - US Manned Orbiting Laboratory

Though never completed, the US Manned Orbiting Laboratory (MOL) played a significant role in shaping the evolution of military space programs in the United States, ultimately influencing the development of the US Space Force decades later. n the context of the US Space Force, the MOL's influence is evident in the organization's mission to secure and protect U.S. interests in space. While the Space Force does not involve crewed missions like the MOL once envisioned, it inherits the legacy of viewing space as a critical national security and military operations domain. The Space Force is charged with safeguarding the U.S. satellite infrastructure that serves both military and civilian purposes, much as the MOL was intended to enhance U.S. military capabilities through space-based reconnaissance.

While the Manned Orbiting Laboratory never flew, its strategic and technological objectives impacted U.S. military space policy. The MOL was an early recognition of the military potential of space. Though its role was eventually overtaken by satellite technology, its emphasis on space-based reconnaissance and national security directly influenced the development of the US Space Force. Today, the Space Force carries forward the vision of defending U.S. interests in space, which was in many ways pioneered by programs like MOL during the Cold War.

In the midst of the Cold War during the mid-1950s, the United States Air Force (USAF) was acutely aware of the Soviet Union's advancing military and industrial capabilities. This concern prompted the USAF to initiate covert reconnaissance efforts, beginning with the U-2 spy plane overflights of Soviet territory in 1956. These missions successfully captured detailed images of approximately fifteen percent of the Soviet Union, achieving resolutions of up to 0.61 meters. However, the program abruptly ended in 1960 when a U-2 aircraft was shot down, leaving a significant gap in American espionage capabilities. In response, the USAF sought to develop alternative means of intelligence gathering, turning its attention to space-based solutions.

The Soviet Union's launch of Sputnik 1 on October 4, 1957, sent shockwaves through the American public and government alike, shattering the prevailing assumption of American technological supremacy. This event not only underscored the feasibility of satellite technology but also highlighted the strategic advantage such capabilities could confer. The absence of any Soviet protest against satellite overflights implicitly acknowledged the legality of such endeavors, thereby removing a significant diplomatic barrier to the deployment of American spy satellites.

Amidst this backdrop, the USAF Wright Air Development Center published a pioneering paper in July 1957 that explored the concept of a space station equipped with telescopes and other observation instruments. This vision laid the groundwork for the USAF's broader satellite ambitions, encapsulated in the WS-117L program initiated in 1956. WS-117L comprised three key components: SAMOS, a dedicated spy satellite; Corona, an experimental initiative aimed at developing the necessary technology; and MIDAS, an early warning system designed to detect incoming missile threats.

Despite the promising developments, the United States faced internal challenges regarding the management and oversight of its burgeoning space programs. In August 1958, President Dwight D. Eisenhower made a strategic decision to transfer responsibility for most forms of human spaceflight to the newly established National Aeronautics and Space Administration (NASA). Deputy Secretary of Defense Donald A. Quarles reallocated $53.8 million, equivalent to approximately $436 million in 2023 dollars, from USAF space projects to NASA. This move left the USAF with only a handful of programs directly impacting military operations, including the Boeing X-20 Dyna-Soar—a delta-wing, rocket-propelled glider envisioned for reconnaissance missions.

Undeterred by the transfer of many space initiatives to NASA, the USAF continued to pursue its own space ambitions. In March 1959, General Thomas D. White, the Chief of Staff of the Air Force, tasked the USAF Director of Development Planning with formulating a long-term space program. This directive identified several projects, notably a "manned orbital laboratory," which would eventually evolve into the Manned Orbiting Laboratory (MOL).

On September 1, 1959, the USAF Air Research and Development Command (ARDC) issued a request to the Aeronautical Systems Division (ASD) at Wright-Patterson Air Force Base for a comprehensive study of a military test space station (MTSS). The ASD solicited input on potential experiments suitable for such a facility, receiving 125 proposals. Subsequently, a request for proposal (RFP) was issued on February 19, 1960, attracting responses from twelve firms. By August 15 of that year, five major contractors—General Electric, Lockheed Aircraft, Glenn L. Martin Company, McDonnell Aircraft Corporation, and General Dynamics—were awarded a combined total of $574,999 (approximately $4.49 million in 2023 dollars) to study the MTSS. Their preliminary reports, submitted in January 1961, laid the foundation for the MOL program, although initial funding requests in August 1961 were unsuccessful.

Parallel to these developments, the USAF was advancing its own spacecraft designs. The Boeing X-20 Dyna-Soar was slated for its first piloted suborbital flight in April 1965, followed by an orbital mission in April 1966, both to be launched by a Titan I booster. However, a memorandum dated February 22, 1962, from the Secretary of Defense, Robert McNamara, directed a significant shift in strategy. To expedite the program and reduce costs, the decision was made to forgo the suborbital testing phase, opting instead to utilize a more powerful Titan III booster for Dyna-Soar launches.

This same memorandum implicitly approved the development of a military space station, now rebranded as the Military Orbital Development System (MODS). By May 1962, a proposed system package plan had been formulated, designated as Program 287 for tracking purposes. MODS was envisioned to include a space station capable of sustaining a crew of four in a shirt-sleeve environment for up to thirty days, a modified version of NASA's Gemini spacecraft known as Blue Gemini for crew transport, and the Titan III launch vehicle. On August 25, 1962, Eugene Zuckert, Deputy Secretary of Defense, formally instructed General Bernard Adolph Schriever, the director of the MOL program, to advance the project. The nomenclature "Manned Orbiting Laboratory" was carefully chosen to maintain plausible deniability, avoiding the overt military connotations of terms like "space station."

As the program gained momentum, Zuckert submitted detailed proposals to Secretary McNamara on November 9, 1962, requesting substantial funding for MODS and Blue Gemini for the fiscal year 1964. Although Project Gemini was initially a NASA endeavor, its association with national security prompted McNamara to consider transferring full control to the Department of Defense. However, after negotiations, an agreement was reached in January 1963 for a collaborative approach between NASA and the Department of Defense, ensuring that both agencies could contribute to the project's success.

In early 1963, McNamara initiated a review to determine whether the Dyna-Soar project possessed unique military capabilities that could not be achieved through the Gemini program. By November 14, 1963, Harold Brown, Director of Defense Research and Engineering, recommended the development of a four-man space station. This station would be launched separately, with astronauts transported via Gemini spacecraft on rotating thirty-day missions, supported by resupply missions every four months to replenish consumables.

The culmination of these efforts was the official announcement of the Manned Orbiting Laboratory program on December 10, 1963. The press release presented MOL to the public as an inhabited platform intended to demonstrate the utility of human presence in space for military purposes. However, the true nature of its reconnaissance mission remained classified, continuing the tradition of semi-secret military space endeavors akin to the earlier Corona spy satellite program. Seventeen astronauts were selected for the MOL program, among them Major Robert H. Lawrence Jr., the first African-American astronaut, symbolizing a progressive step in the integration of diverse personnel within the space program.

The MOL program was spearheaded by General Bernard Adolph Schriever from 1962 to 1966, under the prime contractor McDonnell Aircraft Corporation, with the laboratory itself constructed by the Douglas Aircraft Company. The Gemini B spacecraft, derived from NASA's Gemini but modified to include a circular hatch through the heat shield, facilitated movement between the spacecraft and the laboratory module. Launches were to be conducted from the Vandenberg Space Launch Complex 6 (SLC-6), specifically developed to enable polar orbits essential for reconnaissance missions.

Despite its ambitious design and strategic significance, the MOL program operated under tight security constraints imposed by the Kennedy administration, which sought to mitigate Soviet sensitivities regarding espionage activities in space. The covert nature of MOL meant that, although it had a public facade, its primary mission remained hidden until President Jimmy Carter publicly acknowledged the existence of spy satellites in 1978.

On December 16, 1963, USAF Headquarters issued a directive to General Bernard Adolph Schriever, instructing him to submit a comprehensive development plan for the Manned Orbiting Laboratory (MOL). This mandate initiated a period of intense planning and investment, with approximately six million dollars—equivalent to forty-five million in 2023—allocated to preliminary studies. By September 1964, the majority of these initial studies had been completed. McDonnell Aircraft Corporation undertook a detailed examination of the Gemini B spacecraft, while Martin Marietta focused on the Titan III booster, and Eastman Kodak concentrated on developing advanced camera optics essential for satellite reconnaissance equipment. Concurrently, other critical subsystems of the MOL, including environmental control, electrical power, navigation, attitude control stabilization, guidance, communications, and radar systems, were rigorously analyzed to ensure the laboratory's operational viability.

Brockway McMillan, the Under Secretary of the Air Force and Director of the National Reconnaissance Office (NRO), recognized the strategic importance of MOL's reconnaissance capabilities. He tasked Major General Robert Evans Greer, director of NRO Program A, with conducting an in-depth evaluation of MOL's potential for intelligence gathering. These studies incurred significant expenses, totaling over three million dollars, with the Gemini B spacecraft analysis alone accounting for nearly nine million in today's dollars. The findings underscored the MOL's capacity to enhance America's espionage efforts from space, reinforcing its strategic value.

In January 1965, the USAF expanded its search for expertise by issuing a request for proposals (RFP) to twenty firms for further design studies of the MOL. By the end of February, Boeing, Douglas, General Electric, and Lockheed had been selected to carry out these essential design phases. To maintain operational secrecy, the NRO's activities related to MOL were codified under the designation "Dorian," later renamed KH-10 Dorian in February 1969. Recognizing the necessity of concealing its true mission, MOL incorporated "white" experiments to serve as a public cover. Under the leadership of Colonel William Brady, an Experiments Working Group meticulously reviewed approximately four hundred proposed experiments from various agencies. This extensive review process culminated in a consolidated selection of twelve primary and eighteen secondary experiments, documented in a comprehensive 499-page report released on April 1, 1964. While the laboratory's primary objective was reconnaissance, the program also aimed to demonstrate that astronauts could perform militarily useful tasks in a controlled, shirt-sleeve environment for missions lasting up to thirty days. Innovations such as foot restraints were implemented to secure astronauts at their workstations, a technique that would later be adopted for the International Space Station (ISS).

The strategic blueprint for MOL advocated the use of the Gemini B spacecraft paired with the Titan III booster. The program envisioned six flights—one uncrewed and five crewed—with the inaugural mission slated for 1966. The projected cost of this ambitious endeavor was estimated at approximately 1.653 billion dollars, equivalent to twelve billion in 2023. Donald Hornig, the Science Advisor to President Lyndon B. Johnson, reviewed the USAF's proposal and acknowledged that, for the sophisticated reconnaissance missions envisaged, a human-operated system held distinct advantages over automated alternatives. However, he also noted that with sufficient investment, the performance gap between manned and unmanned systems could potentially be narrowed. While the legality of satellite overflights had been tacitly accepted following the Sputnik crisis, the introduction of a crewed space station posed new diplomatic challenges. Nevertheless, Secretary of State Dean Rusk was confident that these could be effectively managed.

The question of whether the enhanced capabilities of MOL justified its substantial costs remained a topic of debate. Admiral William Raborn, Director of Central Intelligence, concurred that the potential benefits might warrant the investment. Taking this assessment to heart, Secretary of Defense Robert McNamara presented the proposal to President Lyndon Johnson on August 24, 1965. The following day, Johnson formally approved the MOL program, publicly announcing its initiation during a press conference.

In January 1965, Brigadier General Harry L. Evans was appointed by Schriever as his deputy for MOL. Evans, who had previously managed the Corona program and supervised other satellite initiatives such as SAMOS, MIDAS, and SAINT, also assumed the role of Special Assistant for MOL to Deputy Secretary of Defense Eugene Zuckert. In this capacity, he served as the primary liaison between MOL and other governmental agencies, including NASA, ensuring seamless coordination and integration of efforts.

Following Johnson's announcement, MOL was designated Program 632A. The USAF officially named Schriever as the MOL director and Evans as the vice director, overseeing the MOL staff at the Pentagon. Brigadier General Russell A. Berg was appointed as deputy director, responsible for the MOL staff at the Los Angeles Air Force Station in El Segundo, California. The establishment of the MOL System Program Office (SPO) in March 1964, under Brigadier General Joseph S. Bleymaier of the AFSC Space Systems Division (SSD), marked a significant organizational milestone. By August 1965, the MOL team had expanded to include forty-two military personnel and twenty-three civilians, reflecting the program's growing complexity and scope.

In August 1966, Schriever retired from the Air Force, and Major General James Ferguson succeeded him as both head of the AFSC and MOL Program Director. Similarly, Brigadier General Harry L. Evans retired on March 27, 1968, and was replaced by Major General James T. Stewart, ensuring continued leadership and expertise within the MOL program. Throughout its development, MOL utilized mockups to refine its design, facilitating iterative improvements and ensuring the laboratory's readiness for operational deployment.

A critical aspect of MOL's financial management involved maintaining the program's secrecy. On November 4, 1965, Schriever and Alexander H. Flax, Director of the NRO, formalized an agreement outlining MOL's black financial procedures. Under this agreement, the Deputy Director of MOL would submit black budget cost estimates to the NRO Controller, who had the authority to allocate NRO funds accordingly. This was complemented by the MOL White Financial Procedures Agreement, approved in December 1965 and signed by Leonard Marks Jr., the Assistant Secretary of the Air Force for Financial Management and Comptroller. This dual-channel funding mechanism allowed for both classified and unclassified expenditures, ensuring the program's operational integrity without compromising its covert objectives. By September 30, 1965, substantial funding had been released, with twelve million dollars allocated for fiscal year 1965 and fifty million for fiscal year 1966, facilitating the definition phase activities crucial for MOL's progression.

The announcement of MOL contractors included Douglas Aircraft and General Electric. Douglas brought extensive technical and managerial experience from previous projects such as Thor, Genie, and Nike, while General Electric offered expertise in large optical systems and a substantial workforce already cleared for the Dorian mission. On October 17, 1965, a fixed-price contract worth approximately ten and a half million dollars was signed with Douglas, while General Electric secured a contract of nearly five million dollars, predominantly funded through the black budget. The Aerospace Corporation was entrusted with general systems engineering and technical direction, ensuring cohesive integration of the various program components. Douglas, in turn, selected five major subcontractors: Hamilton-Standard for environmental control and life support systems, Collins Radio for communications, Honeywell for attitude control, Pratt & Whitney for electrical power, and IBM for data management. Although Aerospace and the MOL SPO endorsed most of these selections, they expressed reservations about IBM's bid due to its higher cost compared to UNIVAC. Consequently, Douglas opted to allow study contracts to both firms, balancing technical excellence with budgetary constraints.

As the MOL program advanced, the selection and training of its astronauts became a critical component of its development. On June 5, 1961, recognizing the need for highly skilled pilots capable of operating advanced spacecraft, the United States Air Force (USAF) established the Aerospace Research Pilot Course at the USAF Experimental Flight Test Pilot School at Edwards Air Force Base in California. This institution was later renamed the Aerospace Research Pilot School (ARPS) on October 12, 1961. Over the course of two years, from June 1961 to May 1963, ARPS conducted four rigorous classes, with the third cohort receiving specialized instruction on the Dyna-Soar program as part of their curriculum.

Colonel Charles E. "Chuck" Yeager, the esteemed commandant of ARPS, played a pivotal role in shaping the selection criteria for MOL astronauts. He advised General Schriever to limit the pool of candidates to graduates of ARPS, ensuring that only the most capable and well-trained pilots were considered for the program. The selection process was highly selective, with fifteen candidates ultimately chosen and sent to Brooks Air Force Base in San Antonio, Texas, for a week-long medical evaluation in October 1964. These evaluations mirrored those conducted for NASA's astronaut groups, emphasizing the physical and psychological demands of space missions.

The USAF had long collaborated with NASA in the selection of astronauts, having established a selection board for the first three NASA astronaut groups in 1959, 1962, and 1963. Following this precedent, General John P. McConnell, the Chief of Staff of the USAF, informed Schriever that the selection of MOL astronauts would adhere to a similar procedure. In September 1965, a dedicated selection board, chaired by Major General Jerry D. Page, convened to evaluate candidates based on stringent criteria. Announced on September 15, 1965, the selection criteria mandated that candidates be qualified military pilots, graduates of ARPS, serving officers recommended by their commanding officers, and U.S. citizens by birth.

By October 1965, the MOL Policy Committee had decided to designate MOL crew members as "MOL Aerospace Research Pilots" rather than astronauts, a subtle distinction that underscored the program's specialized focus. The first group of eight MOL pilots was publicly announced on November 12, 1965, in a deliberate Friday night news release designed to minimize media attention. This inaugural group included Major Michael J. Adams, Major Albert H. Crews Jr., Lieutenant John L. Finley of the U.S. Navy, Captain Richard E. Lawyer, Captain Lachlan Macleay, Captain Francis G. Neubeck, Major James M. Taylor, and Lieutenant Richard H. Truly of the U.S. Navy. To ensure their continued involvement, Finley and Truly were retained as instructors at Brooks Air Force Base until the official announcement of the MOL program, preventing their return to active naval duty.

In late 1965, the selection process expanded to form a second group of MOL pilots. Unlike the first group, applications were solicited, attracting over five hundred candidates, many of whom also applied to NASA's Astronaut Group 5. The selection board at the MOL Program Office narrowed this pool to twenty-five candidates, who underwent further physical evaluations at Brooks Air Force Base in January and February 1966. From this rigorous selection process, five pilots were chosen and publicly announced on June 17, 1966: Captain Karol J. Bobko, Lieutenant Robert L. Crippen of the U.S. Navy, Captain C. Gordon Fullerton, Captain Henry W. Hartsfield Jr., and Captain Robert F. Overmyer of the U.S. Marine Corps. Notably, Bobko became the first graduate of the United States Air Force Academy to be selected as an astronaut, marking a significant milestone in the integration of academy graduates into the space program.

The third and final group of MOL astronauts was selected on June 30, 1967, following a selection board meeting on May 11, 1967. This group included Major James A. Abrahamson, Lieutenant Colonel Robert T. Herres, Major Robert H. Lawrence Jr., and Major Donald H. Peterson. Robert H. Lawrence Jr. holds the distinction of being the first African-American astronaut, reflecting the program's commitment to diversity and inclusion within its ranks.

Training for MOL astronauts was extensive and multifaceted, designed to prepare them for the unique challenges of space missions. Many of these pilots had nurtured aspirations of space travel since childhood, unaware of the program's true reconnaissance objectives. They were initially informed that MOL was a space laboratory for military experiments, only discovering its covert intelligence mission after their selection. Astronauts were advised to resign if they disagreed with the classified nature of the program. Upon selection, they received security clearances and were briefed on Sensitive Compartmented Information (SCI) programs such as Dorian, Gambit, Talent, and Keyhole, which encompassed various intelligence-gathering operations from spy plane overflights and satellites.

The training regimen for MOL astronauts was divided into several phases. Phase I involved a two-month introductory period, consisting of briefings from NASA and program contractors to familiarize the pilots with the overarching goals and operational frameworks of MOL. Phase II extended over five months at ARPS, where astronauts received technical training on MOL vehicles and operational procedures through classroom instruction, training flights, and sessions on the T-27 space flight simulator. Phase III was an ongoing phase focused on continuous training related to MOL systems and incorporating crew input into their development, a period during which pilots spent the majority of their time. Finally, Phase IV entailed mission-specific training tailored to the objectives and scenarios they would encounter during their missions.

To ensure comprehensive preparedness, specialized simulators were developed for each MOL system, including the Laboratory Module Simulator, Mission Payload Simulator, and Gemini B Procedures Simulator. Zero-gravity training was conducted using a Boeing C-135 Stratolifter, which created reduced-gravity environments, while flotation-egress trainers prepared astronauts for potential splashdowns and water recoveries. Additionally, the program incorporated neutral buoyancy simulations, a technique later adopted by NASA for training International Space Station (ISS) astronauts. Pilots underwent scuba diving training at the U.S. Navy Underwater Swimmers School in Key West, Florida, and participated in simulator training on Buck Island near Saint Thomas in the U.S. Virgin Islands. Water survival training was conducted at the USAF Sea Survival School at Homestead Air Force Base in Florida, complemented by jungle survival training at the Tropical Survival School at Howard Air Force Base in the Panama Canal Zone. In July 1967, astronauts received additional training at the National Photographic Interpretation Center (NPIC) in Washington, D.C., further enhancing their reconnaissance capabilities.

The meticulous selection and rigorous training of MOL astronauts underscored the USAF's commitment to developing a cadre of highly skilled pilots capable of executing complex and classified space missions. These astronauts were not only pioneers of human spaceflight but also integral to the USAF's strategic reconnaissance objectives during the height of the Cold War. The MOL program, with its blend of public scientific endeavors and clandestine intelligence operations, epitomized the intricate balance between transparency and secrecy that characterized much of the space race era.

As the MOL program advanced, the meticulous selection and comprehensive training of its astronauts became paramount to its envisioned success. On June 5, 1961, the United States Air Force (USAF) established the Aerospace Research Pilot Course at the Experimental Flight Test Pilot School located at Edwards Air Force Base in California. This institution, subsequently renamed the Aerospace Research Pilot School (ARPS) on October 12, 1961, was designed to cultivate highly skilled pilots capable of operating the advanced spacecraft integral to the

Dyna-Soar and MOL programs. Between June 1961 and May 1963, ARPS conducted four rigorous classes, with the third cohort receiving specialized instruction tailored to the Dyna-Soar initiative.

Colonel Charles E. "Chuck" Yeager, the esteemed commandant of ARPS, played a pivotal role in shaping the astronaut selection criteria for MOL. Advising General Bernard Adolph Schriever, Yeager recommended restricting the pool of candidates to ARPS graduates, ensuring that only the most capable and thoroughly trained pilots were considered. The selection process was highly selective; fifteen candidates were ultimately chosen and sent to Brooks Air Force Base in San Antonio, Texas, in October 1964 for a week-long medical evaluation. These evaluations mirrored those conducted for NASA's astronaut groups, emphasizing the physical and psychological demands of space missions.

Building on a tradition of collaboration, the USAF had previously worked with NASA to establish selection boards for the first three NASA astronaut groups in 1959, 1962, and 1963. Following this precedent, General John P. McConnell, the Chief of Staff of the USAF, informed Schriever that the selection of MOL astronauts would adhere to a similar rigorous procedure. In September 1965, a dedicated selection board, chaired by Major General Jerry D. Page, convened to evaluate candidates based on stringent criteria. On September 15, 1965, the selection criteria were officially announced, mandating that candidates be qualified military pilots, graduates of ARPS, serving officers recommended by their commanding officers, and U.S. citizens by birth.

By October 1965, the MOL Policy Committee decided to designate MOL crew members as "MOL Aerospace Research Pilots" rather than astronauts, a subtle distinction that underscored the program's specialized focus. The first group of eight MOL pilots was publicly announced on November 12, 1965, through a deliberate Friday night news release designed to minimize media attention. This inaugural group comprised Major Michael J. Adams, Major Albert H. Crews Jr., Lieutenant John L. Finley of the U.S. Navy, Captain Richard E. Lawyer, Captain Lachlan Macleay, Captain Francis G. Neubeck, Major James M. Taylor, and Lieutenant Richard H. Truly of the U.S. Navy. To ensure their continued involvement, Finley and Truly were retained as instructors at Brooks Air Force Base until the official announcement of the MOL program, preventing their return to active naval duty.

In late 1965, the selection process expanded to form a second group of MOL pilots. Unlike the first group, applications were solicited, attracting over five hundred candidates, many of whom also applied to NASA's Astronaut Group 5. The selection board at the MOL Program Office narrowed this pool to twenty-five candidates, who underwent further physical evaluations at Brooks Air Force Base in January and February 1966. From this rigorous selection process, five pilots were chosen and publicly announced on June 17, 1966: Captain Karol J. Bobko, Lieutenant Robert L. Crippen of the U.S. Navy, Captain C. Gordon Fullerton, Captain Henry W. Hartsfield Jr., and Captain Robert F. Overmyer of the U.S. Marine Corps. Notably, Bobko became the first graduate of the United States Air Force Academy to be selected as an astronaut, marking a significant milestone in the integration of academy graduates into the space program.

The third and final group of MOL astronauts was selected on June 30, 1967, following a selection board meeting on May 11, 1967. This group included Major James A. Abrahamson, Lieutenant Colonel Robert T. Herres, Major Robert H. Lawrence Jr., and Major Donald H. Peterson. Robert H. Lawrence Jr. holds the distinction of being the first African-American astronaut, reflecting the program's commitment to diversity and inclusion within its ranks.

Training for MOL astronauts was extensive and multifaceted, designed to prepare them for the unique challenges of space missions. Many of these pilots had nurtured aspirations of space travel since childhood, initially unaware of the program's true reconnaissance objectives. They were informed that MOL was a space laboratory for military experiments, only discovering its covert intelligence mission after their selection. Astronauts were advised to resign if they disagreed with the classified nature of the program. Upon selection, they received security clearances and were briefed on Sensitive Compartmented Information (SCI) programs such as Dorian, Gambit,

Talent, and Keyhole, encompassing various intelligence-gathering operations from spy plane overflights and satellites. Lieutenant Truly, one of the first MOL pilots, remarked on the revelation of a "second space program" that operated parallel to the public NASA efforts, highlighting the program's secrecy.

The training regimen for MOL astronauts was divided into several phases. Phase I involved a two-month introductory period, consisting of briefings from NASA and program contractors to familiarize the pilots with the overarching goals and operational frameworks of MOL. Phase II extended over five months at ARPS, where astronauts received technical training on MOL vehicles and operational procedures through classroom instruction, training flights, and sessions on the T-27 space flight simulator. Phase III was an ongoing phase focused on continuous training related to MOL systems and incorporating crew input into their development, a period during which pilots spent the majority of their time. Finally, Phase IV entailed mission-specific training tailored to the objectives and scenarios they would encounter during their missions.

To ensure comprehensive preparedness, specialized simulators were developed for each MOL system, including the Laboratory Module Simulator, Mission Payload Simulator, and Gemini B Procedures Simulator. Zero-gravity training was conducted using a Boeing C-135 Stratolifter, which created reduced-gravity environments, while flotation-egress trainers prepared astronauts for potential splashdowns and water recoveries. Additionally, the program incorporated neutral buoyancy simulations, a technique later adopted by NASA for training International Space Station (ISS) astronauts. Pilots underwent scuba diving training at the U.S. Navy Underwater Swimmers School in Key West, Florida, and participated in simulator training on Buck Island near Saint Thomas in the U.S. Virgin Islands. Water survival training was conducted at the USAF Sea Survival School at Homestead Air Force Base in Florida, complemented by jungle survival training at the Tropical Survival School at Howard Air Force Base in the Panama Canal Zone. In July 1967, astronauts received additional training at the National Photographic Interpretation Center (NPIC) in Washington, D.C., further enhancing their reconnaissance capabilities.

The planned operations of the MOL program centered on reconnaissance missions that leveraged both automated and human-guided systems. Operating from a regular orbit of 280 kilometers (150 nautical miles), the MOL's main camera featured a circular field of view spanning 2,700 meters (9,000 feet), which could be adjusted to a more detailed view of 1,300 meters (4,200 feet) at top magnification. While this resolution was sufficient for identifying smaller targets, many of the NRO's interests, such as air bases, shipyards, and missile ranges, required more precise imaging. To address this, astronauts would utilize tracking and acquisition telescopes with a circular view approximately 12 kilometers (6.5 nautical miles) across and a resolution of about 9.1 meters (30 feet). The main camera would then focus on the most critical targets, capturing high-resolution images with the most significant aspects of the target centered in the frame. However, due to the optical limitations, image sharpness diminished towards the edges.

Although surveillance targets were pre-programmed and the camera could operate automatically, the presence of astronauts allowed for dynamic decision-making regarding target prioritization. By avoiding cloudy areas and identifying more compelling subjects—such as an open missile silo versus a closed one—the astronauts could optimize the use of limited film resources, a critical constraint given that the film had to be returned to Earth via the compact Gemini B spacecraft. In regions with frequent cloud cover, such as Moscow, the MOL was estimated to be 45 percent more efficient in film usage compared to automated satellite systems, thanks to the astronauts' ability to adapt to changing conditions. In sunnier regions like the Tyuratam missile complex, this efficiency gain was projected to be around 15 percent. The human-guided surveillance offered by MOL was expected to outperform robotic satellites in selective targeting, as evidenced by past KH-7 Gambit satellite missions. For instance, of 159 photographs of the Tyuratam area, only 9 percent captured missiles on their launch pads, and of 77 photographs of missile silos, merely 21 percent showed them with open doors. Analysts identified 60 MOL targets within the complex, with astronauts able to select and photograph the most critical ones with greater resolution than their automated predecessors.

Looking ahead, the USAF envisioned an enhanced version of the MOL space station, known as Block II, slated for the sixth crewed flight in July 1974. Block II was expected to incorporate advanced capabilities such as image transmission and geodetic system targeting. Additionally, astronauts would engage in infrared, multispectral, and ultraviolet astronomy during extended missions on twice-annual flights. Beyond Block II, MOL program managers aspired to construct larger, permanent space facilities. Planning documents depicted ambitious designs for 12-man and 40-man stations, the latter envisioned as a Y-shaped "spaceborne command post" in synchronous orbit. Emphasizing post-attack survivability, the 40-man station was designed to function as a strategic and tactical decision-making hub during a general war, equipped with self-defense capabilities to ensure operational continuity in the event of a conflict.

The flight schedule for the MOL program, as of September 1, 1966, outlined a series of missions intended to demonstrate and refine the program's capabilities. The first mission, scheduled for April 15, 1969, was a Titan IIIM qualification flight simulating an Orbiting Vehicle. This was followed by a second uncrewed Gemini-B/Titan IIIM qualification flight on July 1, 1969, which involved the Gemini-B spacecraft being flown alone without an active laboratory. The program anticipated its first crewed mission on December 15, 1969, with a crew of two—commanded by Major James M. Taylor, potentially accompanied by Major Albert H. Crews Jr.—spending thirty days in orbit. Subsequent missions were planned for April 15, 1970; July 15, 1970; and October 15, 1970, each involving crewed missions lasting between thirty and sixty days. The final planned mission on January 15, 1971, would complete the initial flight schedule.

Central to the MOL program was the Gemini B spacecraft, a modified version of NASA's Gemini spacecraft, which had its origins in 1961 as an evolution of the Mercury spacecraft, initially designated Mercury Mark II. The name "Gemini" reflected its two-man crew configuration. The Gemini B variant featured significant modifications to support the MOL mission profile. Unlike its NASA counterpart, Gemini B included a rear hatch allowing astronauts to enter the MOL space station. Notches were cut into the ejection seat headrests to facilitate access to the hatch, resulting in mirror-image seats instead of identical ones. Additionally, Gemini B was equipped with a larger diameter heat shield to manage the higher energy of reentry from a polar orbit and an increased number of reentry control system thrusters, expanded from four to six. The orbit attitude and maneuvering system (OAMS) was omitted, as capsule orientation for reentry was managed by the forward reentry control system thrusters, and the laboratory module possessed its own reaction control system for orientation.

The Gemini B spacecraft was designed for long-term orbital storage, accommodating up to forty days in space. However, equipment for extended missions was removed to streamline the capsule, which was intended solely for launch and reentry. This design choice meant that the Gemini B had a different cockpit layout and instrumentation compared to the NASA Gemini. Following the tragic Apollo 1 fire in January 1967, which resulted in the deaths of three NASA astronauts during a ground test, the MOL program made significant safety modifications. The program switched from a pure oxygen atmosphere to a helium-oxygen mix within the cabin, while astronauts continued to breathe pure oxygen in their spacesuits during takeoff. This change was already an option in the original design but was now implemented to enhance crew safety.

Four Gemini B spacecraft were ordered from McDonnell Aircraft Corporation, along with a boilerplate aerodynamically similar test article, at a total cost of approximately $168.2 million, equivalent to $1.173 billion in 2023 dollars. In November 1965, NASA agreed to transfer Gemini spacecraft No. 2 and Static Test Article No. 4 to the MOL program. Gemini spacecraft No. 2, which had previously flown in the 1965 Gemini 2 mission, was refurbished to serve as a prototype Gemini B spacecraft, providing invaluable insights into the modifications required for the MOL missions.

The Gemini B was to be launched in tandem with the MOL laboratory module atop a Titan IIIM launch vehicle. Once in orbit, the crew would power down the Gemini B capsule and activate the laboratory module, entering a controlled, shirt-sleeve environment for up to thirty days. After completing their mission, the astronauts would return to the Gemini B capsule, power it back up, detach from the laboratory, and perform reentry procedures. The capsule had an autonomy of approximately fourteen hours once separated from MOL, ensuring a safe return trajectory.

Recovery operations mirrored those of NASA's Project Gemini and Project Apollo, with the Gemini B spacecraft designed to splash down in the Atlantic or Pacific Oceans, where they would be retrieved by Department of Defense (DoD) recovery forces. Although NASA had been developing a paraglider system to enable dry-land landings, the technology was not sufficiently mature by the time of the Gemini missions. In March 1964, NASA attempted to integrate the paraglider with the Gemini B, but the USAF concluded that the system still faced too many technical challenges, ultimately declining the proposal.

Externally, Gemini B bore a striking resemblance to NASA's Gemini spacecraft, yet several critical differences distinguished the two. The most noticeable was the inclusion of a rear hatch for crew entry into the MOL space station. Additionally, Gemini B featured a larger heat shield and an increased number of reentry control thrusters to accommodate the higher reentry velocities from polar orbits. These modifications were essential for the MOL's reconnaissance missions, which required precise orbital parameters and safe reentry capabilities.

Despite the advanced design and strategic importance of the MOL program, it operated under tight security constraints imposed by the Kennedy administration, which sought to mitigate Soviet sensitivities regarding espionage activities in space. The MOL program maintained a public facade as a scientific space laboratory while concealing its true reconnaissance mission, continuing the tradition of semi-secret military space endeavors akin to the earlier Corona spy satellite program. Seventeen astronauts were ultimately selected for the MOL program, including Major Robert H. Lawrence Jr., the first African-American astronaut, symbolizing a progressive step in the integration of diverse personnel within the space program.

As the MOL program progressed, the meticulous planning of its operations became increasingly sophisticated, particularly in the realm of reconnaissance. Operating from a standard orbit of 280 kilometers (150 nautical miles), the MOL's primary camera was engineered to capture expansive fields of view. At a standard magnification, the camera could survey areas spanning 2,700 meters (9,000 feet) across, narrowing to a more detailed 1,300 meters (4,200 feet) at maximum magnification. While this resolution was sufficient for identifying smaller targets, many of the National Reconnaissance Office's (NRO) key interests, such as extensive air bases, sprawling shipyards, and intricate missile ranges, demanded more precise imaging capabilities. To address this, astronauts were equipped with tracking and acquisition telescopes offering a circular view approximately 12 kilometers (6.5 nautical miles) wide and a resolution of about 9.1 meters (30 feet). These telescopes enabled astronauts to identify and prioritize critical targets, which the main camera would then capture with high resolution, ensuring that the most significant aspects of each target were centered in the images. However, due to optical limitations, the sharpness of these images diminished towards the edges, necessitating careful positioning by the crew.

While the surveillance targets were pre-programmed and the main camera could operate autonomously, the presence of astronauts provided a crucial advantage in dynamically prioritizing targets. By actively avoiding areas obscured by cloud cover and identifying more strategically important subjects—such as an open missile silo versus a closed one—the astronauts could optimize the use of limited film resources. This selective targeting was particularly vital given that the film had to be physically returned to Earth via the compact Gemini B spacecraft. In regions with frequent cloud cover, such as Moscow, the MOL's ability to adapt in real-time was estimated to enhance film usage efficiency by 45 percent compared to automated satellite systems. In sunnier regions like the Tyuratam missile complex, the efficiency gain was projected to be around 15 percent. Historical data from previous KH-7 Gambit satellite missions underscored the superiority of human-guided surveillance. For instance, out of 159 photographs of the Tyuratam area, only 9 percent captured missiles on their launch pads, and of 77 photographs of missile silos, merely 21 percent showed them with open doors. In contrast, analysts identified 60 MOL targets within the complex, with astronauts capable of selecting and photographing the most critical ones with greater resolution and relevance than their automated counterparts.

Looking forward, the USAF envisioned an enhanced iteration of the MOL space station, designated Block II, expected to be operational by the sixth crewed flight in July 1974. Block II was projected to incorporate advanced capabilities such as image transmission and geodetic system targeting, significantly enhancing the laboratory's reconnaissance effectiveness. Additionally, astronauts aboard Block II would engage in infrared, multispectral, and ultraviolet astronomy during extended missions, scheduled to occur on twice-annual flights. Beyond Block II, MOL program managers aspired to develop larger, permanent space facilities. Planning documents envisioned 12-man and 40-man stations, with the latter designed as a Y-shaped "spaceborne command post" in synchronous orbit. Emphasizing post-attack survivability, the 40-man station was intended to function as a strategic and tactical decision-making hub during a general war, equipped with self-defense capabilities to ensure operational continuity amidst conflict.

The flight schedule for the MOL program, as of September 1, 1966, outlined a series of missions intended to demonstrate and refine the program's capabilities. The inaugural mission, scheduled for April 15, 1969, was a Titan IIIM qualification flight simulating an Orbiting Vehicle. This was followed by a second uncrewed Gemini-B/Titan IIIM qualification flight on July 1, 1969, which involved the Gemini-B spacecraft being flown alone without an active laboratory. The first crewed mission was anticipated on December 15, 1969, with a crew of two—commanded by Major James M. Taylor, potentially accompanied by Major Albert H. Crews Jr.—spending thirty days in orbit. Subsequent missions were planned for April 15, 1970; July 15, 1970; and October 15, 1970, each involving crewed missions lasting between thirty and sixty days. The final mission on January 15, 1971, was intended to complete the initial flight schedule, culminating in a comprehensive demonstration of MOL's operational capabilities.

Central to the MOL program was the Gemini B spacecraft, a specialized variant of NASA's Gemini spacecraft, which itself had evolved from the Mercury program in 1961. Originally designated Mercury Mark II, the Gemini name was chosen to reflect its two-man crew configuration. The Gemini B was extensively modified to support the MOL mission profile. Unlike the standard Gemini, the Gemini B featured a rear hatch integrated into the heat shield, facilitating direct access to the MOL laboratory module through a transfer tunnel. This tunnel ran through the adaptor module, which housed cryogenic hydrogen, helium, and oxygen storage tanks, the environmental control system, fuel cells, and four quad reaction control system thrusters with their respective propellant tanks. The transfer tunnel was the conduit through which astronauts would move between the Gemini B capsule and the laboratory module, ensuring seamless integration between launch, operations, and reentry phases.

The Gemini B's specifications were meticulously designed to support extended missions. The spacecraft was 3.35 meters (11.0 feet) in length and 2.32 meters (7 feet 7 inches) in diameter, with a cabin volume of 2.55 cubic meters (90 cubic feet). It had a gross mass of 1,983 kilograms (4,372 pounds) and was equipped with 16 Newton thrusters and 98 Newton thrusters, providing a reaction control system (RCS) impulse of 283 seconds (2.78 kilometers per second). The electric system was rated at 4 kilowatt-hours (14 megajoules), supported by a 180 ampere-hour (648,000 coulombs) battery. These specifications ensured that the Gemini B could support a crew of two for up to 40 days in a polar orbit, facilitating both reconnaissance operations and laboratory experiments.

The laboratory module, purpose-built for the MOL program, was a substantial addition to the Gemini B spacecraft. Measuring 5.8 meters (19 feet) in length and 3.05 meters (10.0 feet) in diameter, the module provided a habitable volume of 11.3 cubic meters (400 cubic feet). It was divided into two sections without any internal partitions, allowing the crew to move freely between them. The module's octagonal design featured eight bays, each designated for specific functions. In the "upper" half, Bays 1 and 8 served as storage compartments, Bay 2 housed the environmental control system, Bay 3 was allocated for hygiene and waste management, Bay 4 contained the biochemical test console and workstation, Bays 5 and 6 functioned as the airlock, and Bay 7 was equipped with a glovebox for handling liquids. Additionally, a secondary food console was situated below these bays. The "lower" half of the module included a motion chair in Bay 1 to measure crew mass, two performance test panels in Bay 2, environmental control system controls in Bay 3, a physiology test console in Bay 4, an exercise device in Bay 5, two emergency oxygen masks in Bay 6, a view port and instrument panel in Bay 7, and the main spacecraft control station in Bay 8. This intricate layout ensured that the laboratory could support a wide range of experiments and operational tasks during the crew's mission.

Integral to the MOL program were the specially designed spacesuits, tailored to the unique requirements of the Gemini B spacecraft. The confined space within the Gemini B necessitated a more flexible suit than those used by NASA astronauts. Unlike NASA's custom-made sets, the MOL spacesuits were intended to be provided in standard sizes with adjustable elements to accommodate different crew members. In 1964, the USAF solicited design proposals from several manufacturers, including the David Clark Company, International Latex Corporation, B. F. Goodrich, and Hamilton Standard. Both Hamilton Standard and David Clark developed four prototype suits for the MOL, leading to a competitive selection process. In January 1967, a competition was held at Wright-Patterson Air Force Base, culminating in a production contract awarded to Hamilton Standard. Between May 1968 and July 1969, at least seventeen blue MOL MH-7 training suits were delivered, with a single MH-8 flight configuration suit provided in October 1968 for certification testing. The MH-8 suit was designated for use during launch and reentry, ensuring crew safety during these critical phases.

Further advancements in spacesuit technology were pursued to facilitate extravehicular activities (EVAs). In September 1967, a second competition was held to design an EVA suit capable of addressing USAF concerns about crew members potentially slipping their tethers and floating away. This led to the development of an astronaut maneuvering unit (AMU), integrated with the life support system to form an integrated maneuvering and life support system (IMLSS). The design was finalized by October 1968, and a prototype, excluding cover garments, was delivered in March 1969. Although the cover garments were never completed, the IMLSS represented a significant technological innovation aimed at enhancing the safety and functionality of MOL astronauts during spacewalks.

The physical infrastructure supporting the MOL program was equally critical to its success. The primary launch site, initially considered to be Cape Kennedy, was ultimately designated to Vandenberg Air Force Base's Space Launch Complex 6 (SLC-6). The decision to launch from the West Coast was driven by the requirement for polar orbits essential for reconnaissance missions. However, this choice faced significant logistical and political challenges. Launches from Cape Kennedy would necessitate a "dog leg" maneuver to avoid overflying southern Florida, raising safety concerns and complicating mission profiles due to the need for State Department approval for overflights of Cuba. Additionally, the required modifications would reduce the orbital payload capacity, limiting the equipment that could be carried or the mission duration.

The USAF encountered opposition from Florida's media and political representatives, who criticized the decision as a duplicative and costly endeavor, given the existence of NASA's recently completed Space Launch Complex 41 at Cape Canaveral. In response to mounting pressure, the USAF defended the necessity of SLC-6 for polar-orbit missions during congressional hearings in February 1966. Testimonies from NASA's Associate Administrator Robert Seamans and USAF officials like General Bernard Schriever provided a united front, ultimately silencing significant opposition and securing the necessary budget allocations.

Despite initial setbacks in land acquisition due to failed negotiations over purchase prices, the USAF proceeded with the construction of SLC-6 through eminent domain, acquiring substantial acreage from the Sudden Ranch and Scolari Ranch for approximately $9 million, equivalent to $64.6 million in 2023 dollars. Groundbreaking for the new launch complex occurred on March 12, 1966, followed by extensive earthworks and the construction of essential infrastructure, including access roads, water supply pipelines, and a railroad siding. By August 1968, major components of the launch complex, such as the launch control center, segment receipt inspection building, and ready building, were completed and accepted by the USAF.

The design of SLC-6 had reached a stage where construction bids could be solicited. Key elements included a launch pad, umbilical tower, mobile services tower, aerospace ground equipment building, propellant loading and storage systems, launch control center, segment receipt inspection building, ready building, protective clothing building, and complex service building. Among the seven bids received, the contract was awarded to Santa Fe and Stolte of Lancaster, California, for $20.2 million, equivalent to $145 million in 2023 dollars. The United States Army Corps of Engineers oversaw the construction, ensuring that the launch complex met the stringent requirements necessary for MOL's advanced reconnaissance missions.

At the heart of the MOL program's operational design was the seamless integration of the Gemini B spacecraft with the laboratory module. Once launched atop a Titan IIIM launch vehicle, the Gemini B would deploy the laboratory module, creating a controlled, shirt-sleeve environment for the crew. After approximately thirty days of conducting reconnaissance missions and performing scientific experiments, the astronauts would return to the Gemini B capsule, reengage the propulsion systems, and execute reentry procedures. The capsule's autonomy of around fourteen hours post-separation from MOL ensured a safe and controlled descent back to Earth.

Recovery operations mirrored those of NASA's Project Gemini and Project Apollo, with the Gemini B designed to splash down in the Atlantic or Pacific Oceans. The Department of Defense's recovery forces were tasked with retrieving the capsule, ensuring the safe return of the astronauts and the vital reconnaissance film. Although NASA had been developing a paraglider system intended to allow Gemini capsules to glide to dry-land touchdowns, the USAF concluded that the technology was not sufficiently reliable by the time of the Gemini missions and thus opted against its integration into the Gemini B design.

Externally, the Gemini B bore a strong resemblance to NASA's Gemini spacecraft, yet several critical modifications distinguished the two. The most prominent was the rear hatch incorporated into the heat shield, enabling direct access to the laboratory module. Additionally, the Gemini B featured a larger diameter heat shield to accommodate the higher energy reentries from polar orbits and an increased number of reentry control thrusters, expanding from four to six. The omission of an orbit attitude and maneuvering system (OAMS) was compensated by the forward reentry control system thrusters managing capsule orientation and the laboratory module's own reaction control system for maintaining its position and stability.

Designed for long-term orbital storage of up to forty days, the Gemini B's systems were streamlined for launch and reentry, with non-essential long-duration equipment removed to reduce weight and complexity. The cockpit layout and instrumentation were reconfigured to better suit the MOL's operational requirements. Following the Apollo 1 tragedy in January 1967, which resulted in the deaths of three NASA astronauts during a ground test, the MOL program implemented significant safety enhancements. The cabin atmosphere was switched from a pure oxygen environment to a helium-oxygen mix, enhancing safety during takeoff and reentry. Astronauts would breathe pure oxygen in their spacesuits while the cabin was pressurized with helium, a precautionary measure that had been incorporated into the original design.

The procurement of four Gemini B spacecraft from McDonnell Aircraft Corporation, along with a boilerplate aerodynamically similar test article, amounted to a total cost of approximately $168.2 million, equivalent to $1.173 billion in 2023 dollars. In November 1965, NASA transferred Gemini spacecraft No. 2 and Static Test Article No.

4 to the MOL program. Gemini spacecraft No. 2, which had previously participated in the 1965 Gemini 2 mission, was refurbished to serve as a prototype Gemini B, providing invaluable insights into the modifications required for MOL's unique mission profile.

Despite its advanced design and strategic significance, the MOL program operated under stringent security constraints imposed by the Kennedy administration, which aimed to mitigate Soviet sensitivities regarding espionage activities in space. The MOL program maintained a dual identity: publicly presenting itself as a scientific space laboratory for military experiments while covertly executing reconnaissance missions under the classified codename "Dorian." This duality echoed the earlier Corona spy satellite program, blending transparency with secrecy to obscure the program's true objectives from public scrutiny and international observers.

As the MOL program advanced, the meticulous planning of its operations extended beyond the confines of the United States, encompassing strategic international collaborations and infrastructural developments essential for mission success. Recognizing the inherent risks associated with space missions, particularly abort scenarios, the USAF proactively established contingency plans to ensure crew safety. In the event of an abort, the Gemini B spacecraft could potentially descend into the vast expanse of the eastern Pacific Ocean. To prepare for such contingencies, an agreement was reached with Chile on July 26, 1968, permitting the use of Easter Island as a staging area for search and rescue operations. This agreement facilitated the resurfacing of the island's 2,000-meter (6,600 feet) runway, along with the construction of taxiways and parking areas paved with asphalt. Additionally, comprehensive facilities for communications, aircraft maintenance, storage, and accommodation for up to one hundred personnel were established, ensuring that rescue teams could be rapidly deployed and effectively supported in the event of an emergency.

Parallel to these contingency measures, the MOL program necessitated the development of specialized reconnaissance equipment. A pivotal component of this effort was the establishment of the Camera Optical Assembly (COA) facility at Eastman Kodak in Rochester, New York. This state-of-the-art facility featured a new steel frame building and a subterranean masonry structure encompassing 13,120 square meters (141,200 square feet) of test chambers. Constructed at a cost of approximately $32.5 million—equivalent to $240 million in 2023—the laboratory's concealed design ensured that its true purpose remained obscured from public view, maintaining the program's semi-secret status.

In pursuit of operational readiness, the MOL program conducted its inaugural test flight on November 3, 1966, from Cape Canaveral's Space Launch Complex 40. The mission, designated OPS 0855, utilized a Titan IIIC launch vehicle carrying a MOL mockup constructed from a Titan II propellant tank and the refurbished Gemini B spacecraft No. 2. This uncrewed flight marked a significant milestone as it was the first instance of an American spacecraft intended for human spaceflight achieving two spaceflights, albeit without a crew. During the mission, the Gemini B separated for a suborbital reentry while the MOL mockup continued into low Earth orbit, successfully deploying three satellites: OV4-1, OV1-6, and another OV4-1. The simulated laboratory module housed eleven experiments, encompassing micrometeoroid detection, thermal control, fluid dynamics in zero gravity, and other critical scientific investigations. The mission demonstrated the viability of the MOL's systems, with eight out of the eleven experiments achieving success. The Gemini capsule's hatch, integrated into the heat shield, was tested during reentry, and the capsule was subsequently recovered near Ascension Island in the South Atlantic by the USS La Salle after a 33-minute flight. The laboratory mockup remained in orbit until its decay on January 9, 1967, validating the program's design and operational procedures.

Despite these technical achievements, the MOL program faced significant scrutiny on the international stage, particularly in the context of the 1966 Eighteen Nation Committee on Disarmament. Concerns arose regarding the program's alignment with the United Nations General Assembly resolution of October 17, 1963, which advocated for the exploration and use of outer space solely "for the betterment of mankind." The United States sought to reassure the international community by proposing that Soviet officials be permitted to inspect the MOL facilities prior to launch. However, Secretary of Defense Robert McNamara opposed this suggestion on security grounds, citing the classified nature of the program. Public debate within the United States was hindered by MOL's semi-secret status, limiting transparency and fostering speculation. Leonard E. Schwartz, a consultant to the Directorate for Scientific Affairs of the OECD, remarked in 1967 that without full disclosure of the program's capabilities, it was impossible to objectively assess its costs and benefits relative to existing satellite reconnaissance systems like SAMOS and Vela.

Domestically, the MOL program was presented to the public in a manner that emphasized its scientific and military utility without revealing its true reconnaissance mission. The Air Force characterized MOL as "an effective space building block of very substantial potential," capable of supporting "manned military space operations" and acquiring the necessary crews, experience, and equipment to maintain a presence in near-Earth space. Colonel William Brady, leading the MOL Experiments Working Group, articulated in 1965 that MOL would configure and conduct operations that would enable the Air Force to move into the space environment "in an orderly and effective manner."

Amidst these efforts, the Soviet Union accelerated its own military space ambitions, commissioning the development of the Almaz space station. Initiated by chief designer Vladimir Chelomey on October 12, 1964, the Almaz project gained official endorsement and funding following President Lyndon Johnson's announcement of the MOL program on August 25, 1965. The Almaz stations flew as Salyut space stations between 1973 and 1976 before the crewed Almaz program was ultimately canceled in 1978, reflecting the intense competition and parallel developments characterizing the Cold War space race.

The MOL program's trajectory, however, was marred by persistent budgetary constraints and escalating costs. Shortly after Johnson's announcement, the program faced significant budget cuts. In November 1965, Secretary Flax arbitrarily reduced the MOL program's fiscal year 1967 budget by $20 million—equivalent to $148 million in 2023 dollars—down to $374 million (approximately $2.76 billion in 2023). Concurrently, Secretary McNamara sought to limit the program's budget to $150 million (equivalent to $1.076 billion in 2023) for fiscal year 1967, matching the previous year's allocation despite rising costs driven by concurrent military engagements such as the Vietnam War.

As the MOL program entered its engineering development phase in September 1966, discrepancies between USAF cost estimates and contractor projections became apparent. McDonnell Aircraft Corporation requested $205.5 million (equivalent to $1.475 billion in 2023) for a fixed-price plus incentive fee (FPIF) contract to design and build the Gemini B spacecraft, while the USAF had budgeted $147.9 million (equivalent to $1.061 billion in 2023). Similarly, Douglas Aircraft Company sought $815.8 million (equivalent to $5.854 billion in 2023) for the laboratory vehicles, against a USAF budget of $611.3 million (equivalent to $4.386 billion in 2023), and General Electric requested $198 million (equivalent to $1.421 billion in 2023) for work budgeted at $147.3 million (equivalent to $1.057 billion in 2023). In response, the MOL System Program Office (SPO) reopened negotiations for systems not yet under contract and halted the issuance of Dorian clearances to contractor personnel. This strategic maneuver led to reduced pricing from major contractors by December 1966, though Secretary of Defense McNamara continued to impose stringent budgetary limits. By January 1967, the Office of the Secretary of Defense (OSD) directed the MOL SPO to operate within a fiscal year 1968 budget of $430 million (equivalent to $3.085 billion in 2023), $157 million (equivalent to $1.127 billion in 2023) below the MOL SPO's request and $381 million (equivalent to $2.734 billion in 2023) below contractor demands. This necessitated renegotiations of prime contracts, further delaying the program's progress.

Compounding these financial challenges were delays in critical component deliveries. On December 9, 1966, Eastman Kodak informed the MOL program that it could not deliver the optical sensors by the original target date of January 1969 for the first crewed mission in April 1969, requesting a ten-month extension to October 1969. This delay pushed the first crewed mission to January 1970, disrupting the tightly coordinated flight schedule. Despite securing additional funding through reprogrammed funds from other projects and an eventual appropriation of $661 million (equivalent to $4.743 billion in 2023) for fiscal year 1969, the MOL program's projected costs surged to $2.35 billion (equivalent to $17 billion in 2023).

In a final bid to stabilize the program's trajectory, Brigadier General Joseph S. Bleymaier, Deputy Commander of the AFSC Space Systems Division (SSD), in early 1968 sought input from MOL astronauts, who, acutely aware of the program's financial and political precariousness, advocated for the assurance that the first launch would be fully operational. This recommendation led to the cancellation of two uncrewed qualification missions, thereby designating the first crewed mission, rescheduled from August to December 1971, as the inaugural operational flight. This adjustment was formalized through a series of contracts, including a $674.7 million FPIF contract with Douglas Aircraft Corporation and substantial agreements with McDonnell Aircraft Corporation and General Electric, further inflating the program's costs but aligning contractual obligations with revised budgetary constraints.

Integral to the MOL program's operational infrastructure was the establishment of a dedicated computer center, which utilized state-of-the-art computers for design and simulation. This facility was essential for managing the complex data and ensuring the precision required for reconnaissance operations, though it also represented a significant financial and logistical investment within the broader scope of the program.

As the MOL program matured, critical evaluations regarding the necessity of human involvement in space reconnaissance began to surface. Mere months into its development, the program initiated the creation of an automated variant of MOL, which replaced the crew compartment with film reentry vehicles. This shift was prompted by early tests of human reconnaissance from space, notably the Gemini 5 mission in 1965. During this mission, seventeen USAF military experiments were conducted, including the photographing of missile launches from Vandenberg Air Force Base and observations of the White Sands Proving Ground. However, the National Photographic Interpretation Center (NPIC), responsible for reviewing all NASA astronaut photographs of Earth before their public release, determined that the Gemini photographs lacked utility, as they failed to provide precise data on camera targeting.

Despite these initial setbacks, advancements in automated technology sparked concerns within the MOL program that the role of astronauts might become redundant. Astronaut Al Crews expressed his apprehension, stating, "it became obvious that all we were was a backup in case the unmanned reconnaissance system didn't work." Lieutenant Robert L. Crippen, while acknowledging the rapid improvement of automation, maintained that human presence could not be entirely supplanted by machines. This growing debate underscored a fundamental question within the program: the intrinsic value of human judgment and adaptability in space reconnaissance.

In February 1966, General Schriever commissioned a comprehensive report to assess the continued usefulness of humans aboard the MOL space station. Submitted on May 25, 1966, the report concluded that astronauts retained several key advantages over automated systems. It highlighted the superior resolution of the Dorian camera aboard MOL compared to the KH-8 Gambit 3 satellites, which could not achieve the same level of detail. Additionally, the report emphasized the flexibility afforded by human operators, such as the ability to select optimal photographic angles, switch between different types of film (e.g., color and infrared), and adapt to changing conditions like cloud cover. These capabilities were particularly valuable for identifying camouflaged targets and adjusting missions in real-time to maximize the utility of limited film resources—a critical constraint given that the film had to be physically returned to Earth via the compact Gemini B spacecraft.

Moreover, the report noted that the MOL's ability to alter its orbit from a standard 280 kilometers (150 nautical miles) to a range between 370 and 560 kilometers (200 to 300 nautical miles) provided a comprehensive view of the Soviet Union, enhancing the effectiveness of reconnaissance missions. The authors posited that a crewed MOL would be inherently more resilient and capable of "self-healing" than an unmanned counterpart. Drawing from experiences in Projects Mercury, Gemini, and the X-15, they observed that crew initiative, innovation, and the capacity to improvise were often decisive factors in mission success or failure. Consequently, the report advocated for the continued inclusion of astronauts in the MOL program, asserting that human-guided missions would likely outperform their automated predecessors in both reliability and operational efficiency.

Despite these compelling arguments, debate persisted regarding the overall value of the MOL program, particularly in relation to the high-resolution imaging it promised versus the capabilities of existing and emerging satellite technologies. Following incidents such as the Liberty incident in June 1967 and the Pueblo incident in January 1968, there was an intensified focus on satellite-based intelligence gathering. The Director of Central Intelligence, Richard M. Helms, commissioned a report in May 1968 to evaluate the merits of Very High Resolution (VHR) imaging. The report concluded that while VHR would enhance the identification of smaller objects and provide deeper insights into Soviet industrial capacities and operational procedures, it would not significantly alter assessments of Soviet technical capabilities or the size and deployment of its military forces. The determination of whether the benefits of VHR justified the substantial costs remained inconclusive, yet by 1968, the USAF decided that the automated system's prolonged development time and uncertain performance metrics necessitated the immediate deployment of astronauts for the initial MOL missions.

Amidst these ongoing evaluations, the political landscape underwent significant shifts that would ultimately determine the fate of the MOL program. On January 20, 1969, Richard Nixon was inaugurated as the President of the United States. In his inaugural days, Nixon instructed Robert Mayo, the newly appointed Director of the Bureau of the Budget, and Melvin Laird, the Secretary of Defense, to identify and implement defense spending cuts. The MOL program, with its substantial budget and overlapping objectives with other intelligence programs such as the CIA's Hexagon, quickly emerged as a prime target for budget reductions. An influential article in the Washington Monthly, titled "How The Pentagon Can Save $9 Billion" by former Department of Defense employee Robert S. Benson, criticized the MOL as an exorbitantly expensive initiative that "ought to rank dead last on any rational scale of national priorities."

Competing for funds with the CIA's Hexagon program, MOL faced scrutiny despite their differing objectives. Both programs utilized Titan III launchers and demanded comparable financial investments, while the NRO's already operational and less costly Gambit 3 satellites continued to deliver consistent reconnaissance capabilities. As the MOL program's satellites promised higher resolution imagery, the escalating costs and delays made it increasingly difficult to justify their continued funding.

Amid these budgetary constraints, internal discussions within the Department of Defense revealed a divide regarding the program's viability. Major General James T. Stewart, acting as the MOL Program Director following Brigadier General Harry L. Evans' retirement, briefed the new Deputy Secretary of Defense, David Packard, on the strategic importance of MOL's high-resolution capabilities. Stewart advocated for MOL as the premier path to achieving VHR at the earliest possible date, emphasizing its superiority over existing automated systems. Conversely, Robert Mayo, representing the Bureau of the Budget, contended that the resolution provided by the Gambit 3 satellites was adequate, and proposed the cancellation of both the MOL and Hexagon programs. He argued that the MOL missions, estimated to cost $150 million (equivalent to $956 million in 2023), were not justified when compared to the significantly cheaper Gambit 3 launches at $23 million (equivalent to $147 million in 2023). Mayo's position was rooted in the belief that the benefits of VHR were not substantial enough to warrant the exorbitant costs, especially given the advancements in automated reconnaissance technologies.

Despite efforts by Packard and CIA Director Helms to advocate for MOL's continuation, presenting arguments that Hexagon's broader surveillance capabilities were crucial for arms control—an area of heightened importance to President Nixon—the Bureau of the Budget maintained its stance. In May 1968, a reconnaissance advisory panel led by Edwin Land recommended the cancellation of the MOL program, suggesting that Dorian's technology be utilized in an unmanned form instead. This recommendation, combined with the ongoing budgetary pressures, led to a decisive shift in policy.

On June 7, 1969, Major General James T. Stewart ordered the cessation of all work on the Gemini B spacecraft, the Titan IIIM launch vehicle, and MOL-specific spacesuits. Simultaneously, all other contracts related to the MOL program were either canceled or curtailed. The official announcement of the MOL program's termination was made on June 10, 1969. The cancellation sent shockwaves through the aerospace community and resulted in the loss of thousands of jobs, including 500 at Eastman Kodak alone, which had been heavily invested in the program's optical systems.

In hindsight, the MOL program's cancellation was a confluence of escalating costs, shifting political priorities, and the rapid advancement of automated reconnaissance technologies that rendered human-operated missions less essential. Had the program continued as scheduled, MOL would have been the world's first space station, offering unprecedented reconnaissance capabilities through its high-resolution imaging systems. However, the financial burdens and the perception that unmanned systems could achieve similar or superior outcomes without the added complexities and costs associated with human spaceflight led to its termination.

Post-cancellation reflections by key figures underscored the contentious nature of the decision. General Lew Allen, who later became Chief of Staff of the Air Force, opined that the decision to pursue an unmanned MOL introduced design changes that ultimately contributed to the program's downfall. NRO Director John McLucas criticized the program's budgetary inefficiency, noting that "little tangible [was] to show for the $1.4 billion spent." Additionally, some within the MOL astronaut corps, including Major James A. Abrahamson, later acknowledged that their advocacy for launching fully operational missions may have inadvertently hastened the program's cancellation by complicating its budgetary and technical feasibility.

Al Crews, another prominent MOL astronaut, believed that automated systems were likely superior to crewed missions, a sentiment that resonated as high-resolution photographs from the Gambit 3 satellites began to rival those projected for MOL. This technological parity diminished the perceived necessity of maintaining a human presence in the MOL space station, reinforcing the arguments for its cancellation.

The MOL program's abrupt end left a significant void in the USAF's space reconnaissance efforts. While the program never achieved operational status, its ambitious vision and the technological advancements it spurred continued to influence subsequent space initiatives. The lessons learned from MOL's development—ranging from the complexities of integrating human operations with advanced reconnaissance technology to the challenges of managing large-scale, semi-secret military programs—provided valuable insights that shaped future collaborations between military and civilian space agencies.

As the MOL program progressed, critical evaluations regarding the necessity of human involvement in space reconnaissance began to surface. Mere months into its development, the program initiated the creation of an automated variant of MOL, which replaced the crew compartment with film reentry vehicles. This shift was prompted by early tests of human reconnaissance from space, notably the Gemini 5 mission in 1965. During this mission, seventeen USAF military experiments were conducted, including the photographing of missile launches from Vandenberg Air Force Base and observations of the White Sands Proving Ground. However, the National Photographic Interpretation Center (NPIC), responsible for reviewing all NASA astronaut photographs of Earth before their public release, determined that the Gemini photographs lacked utility, as they failed to provide precise data on camera targeting.

Despite these initial setbacks, advancements in automated technology sparked concerns within the MOL program that the role of astronauts might become redundant. Astronaut Al Crews expressed his apprehension, stating, "it became obvious that all we were was a backup in case the unmanned reconnaissance system didn't work." Lieutenant Robert L. Crippen, while acknowledging the rapid improvement of automation, maintained that human presence could not be entirely supplanted by machines. This growing debate underscored a fundamental question within the program: the intrinsic value of human judgment and adaptability in space reconnaissance.

In February 1966, General Schriever commissioned a comprehensive report to assess the continued usefulness of humans aboard the MOL space station. Submitted on May 25, 1966, the report concluded that astronauts retained several key advantages over automated systems. It highlighted the superior resolution of the Dorian camera aboard MOL compared to the KH-8 Gambit 3 satellites, which could not achieve the same level of detail. Additionally, the report emphasized the flexibility afforded by human operators, such as the ability to select optimal photographic angles, switch between different types of film (e.g., color and infrared), and adapt to changing conditions like cloud cover. These capabilities were particularly valuable for identifying camouflaged targets and adjusting missions in real-time to maximize the utility of limited film resources—a critical constraint given that the film had to be physically returned to Earth via the compact Gemini B spacecraft.

Moreover, the report noted that the MOL's ability to alter its orbit from a standard 280 kilometers (150 nautical miles) to a range between 370 and 560 kilometers (200 to 300 nautical miles) provided a comprehensive view of the Soviet Union, enhancing the effectiveness of reconnaissance missions. The authors posited that a crewed MOL would be inherently more resilient and capable of "self-healing" than an unmanned counterpart. Drawing from experiences in Projects Mercury, Gemini, and the X-15, they observed that crew initiative, innovation, and the capacity to improvise were often decisive factors in mission success or failure. Consequently, the report advocated for the continued inclusion of astronauts in the MOL program, asserting that human-guided missions would likely outperform their automated predecessors in both reliability and operational efficiency.

Despite these compelling arguments, debate persisted regarding the overall value of the MOL program, particularly in relation to the high-resolution imaging it promised versus the capabilities of existing and emerging satellite technologies. Following incidents such as the Liberty incident in June 1967 and the Pueblo incident in January 1968, there was an intensified focus on satellite-based intelligence gathering. The Director of Central Intelligence, Richard M. Helms, commissioned a report in May 1968 to evaluate the merits of Very High Resolution (VHR) imaging. The report concluded that while VHR would enhance the identification of smaller objects and provide deeper insights into Soviet industrial capacities and operational procedures, it would not significantly alter assessments of Soviet technical capabilities or the size and deployment of its military forces. The determination of whether the benefits of VHR justified the substantial costs remained inconclusive, yet by 1968, the USAF decided that the automated system's prolonged development time and uncertain performance metrics necessitated the immediate deployment of astronauts for the initial MOL missions.

Amidst these ongoing evaluations, the political landscape underwent significant shifts that would ultimately determine the fate of the MOL program. On January 20, 1969, Richard Nixon was inaugurated as the President of the United States. In his inaugural days, Nixon instructed Robert Mayo, the newly appointed Director of the Bureau of the Budget, and Melvin Laird, the Secretary of Defense, to identify and implement defense spending cuts. The MOL program, with its substantial budget and overlapping objectives with other intelligence programs such as the CIA's Hexagon, quickly emerged as a prime target for budget reductions. An influential article in the Washington Monthly, titled "How The Pentagon Can Save $9 Billion" by former Department of Defense employee Robert S. Benson, criticized the MOL as an exorbitantly expensive initiative that "ought to rank dead last on any rational scale of national priorities."

Competing for funds with the CIA's Hexagon program, MOL faced scrutiny despite their differing objectives. Both programs utilized Titan III launchers and demanded comparable financial investments, while the NRO's already operational and less costly Gambit 3 satellites continued to deliver consistent reconnaissance capabilities.

As the MOL program's satellites promised higher resolution imagery, the escalating costs and delays made it increasingly difficult to justify their continued funding.

Amid these budgetary constraints, internal discussions within the Department of Defense revealed a divide regarding the program's viability. Major General James T. Stewart, acting as the MOL Program Director following Brigadier General Harry L. Evans' retirement, briefed the new Deputy Secretary of Defense, David Packard, on the strategic importance of MOL's high-resolution capabilities. Stewart advocated for MOL as the premier path to achieving VHR at the earliest possible date, emphasizing its superiority over existing automated systems. Conversely, Robert Mayo, representing the Bureau of the Budget, contended that the resolution provided by the Gambit 3 satellites was adequate, and proposed the cancellation of both the MOL and Hexagon programs. He argued that the MOL missions, estimated to cost $150 million (equivalent to $956 million in 2023), were not justified when compared to the significantly cheaper Gambit 3 launches at $23 million (equivalent to $147 million in 2023). Mayo's position was rooted in the belief that the benefits of VHR were not substantial enough to warrant the exorbitant costs, especially given the advancements in automated reconnaissance technologies.

Despite efforts by Packard and CIA Director Helms to advocate for MOL's continuation, presenting arguments that Hexagon's broader surveillance capabilities were crucial for arms control—an area of heightened importance to President Nixon—the Bureau of the Budget maintained its stance. In May 1968, a reconnaissance advisory panel led by Edwin Land recommended the cancellation of the MOL program, suggesting that Dorian's technology be utilized in an unmanned form instead. This recommendation, combined with the ongoing budgetary pressures, led to a decisive shift in policy.

On June 7, 1969, Major General James T. Stewart ordered the cessation of all work on the Gemini B spacecraft, the Titan IIIM launch vehicle, and MOL-specific spacesuits. Simultaneously, all other contracts related to the MOL program were either canceled or curtailed. The official announcement of the MOL program's termination was made on June 10, 1969. The cancellation sent shockwaves through the aerospace community and resulted in the loss of thousands of jobs, including 500 at Eastman Kodak alone, which had been heavily invested in the program's optical systems.

In hindsight, the MOL program's cancellation was a confluence of escalating costs, shifting political priorities, and the rapid advancement of automated reconnaissance technologies that rendered human-operated missions less essential. Had the program continued as scheduled, MOL would have been the world's first space station, offering unprecedented reconnaissance capabilities through its high-resolution imaging systems. However, the financial burdens and the perception that unmanned systems could achieve similar or superior outcomes without the added complexities and costs associated with human spaceflight led to its termination.

Post-cancellation reflections by key figures underscored the contentious nature of the decision. General Lew Allen, who later became Chief of Staff of the Air Force, opined that the decision to pursue an unmanned MOL introduced design changes that ultimately contributed to the program's downfall. NRO Director John McLucas criticized the program's budgetary inefficiency, noting that "little tangible [was] to show for the $1.4 billion spent." Additionally, some within the MOL astronaut corps, including Major James A. Abrahamson, later acknowledged that their advocacy for launching fully operational missions may have inadvertently hastened the program's cancellation by complicating its budgetary and technical feasibility.

Al Crews, another prominent MOL astronaut, believed that automated systems were likely superior to crewed missions, a sentiment that resonated as high-resolution photographs from the Gambit 3 satellites began to rival those projected for MOL. This technological parity diminished the perceived necessity of maintaining a human presence in the MOL space station, reinforcing the arguments for its cancellation.

The MOL program's abrupt end left a significant void in the USAF's space reconnaissance efforts. While the program never achieved operational status, its ambitious vision and the technological advancements it spurred continued to influence subsequent space initiatives. The lessons learned from MOL's development—ranging from the complexities of integrating human operations with advanced reconnaissance technology to the challenges of managing large-scale, semi-secret military programs—provided valuable insights that shaped future collaborations between military and civilian space agencies.

Integral to the MOL program's operational infrastructure was the establishment of a dedicated computer center, which utilized state-of-the-art computers for design and simulation. This facility was essential for managing the complex data and ensuring the precision required for reconnaissance operations, though it also represented a significant financial and logistical investment within the broader scope of the program.

Despite the persistent financial and technical hurdles, the MOL program embodied the United States' unwavering commitment to maintaining a strategic advantage in space during the height of the Cold War. The combination of advanced technological developments, rigorous astronaut training, and strategic international collaborations underscored the program's significance within the broader narrative of the Space Race. However, escalating costs, shifting political priorities, and the inherent challenges of integrating military objectives with the demands of human spaceflight ultimately culminated in the program's cancellation in the early 1970s. Nevertheless, the MOL program's legacy endured, influencing subsequent space initiatives and highlighting the complex interplay between military ambition and the pursuit of human exploration beyond Earth's atmosphere.

Following the cancellation, a committee was formed to manage the disposal of MOL's assets, valued at approximately $12.5 million (equivalent to $80 million in 2023). Key components such as the Acquisition and Tracking System, Mission Development Simulator, Laboratory Module Simulator, and Mission Simulator were transferred to NASA by the end of 1973. The MOL Program Office at the Pentagon was officially closed on February 15, 1970, followed by the office in Los Angeles on September 30, 1970. Brigadier General Allen, the Director of Space Systems, became the primary contact for terminating contracts, although negotiations with contractors like Aerojet, McDonnell Douglas, and United Technologies Corporation (UTC) continued until June 1973. While some contracts were settled with minimal claims, others, particularly with Douglas and UTC, remained unresolved due to subcontractor disputes and the complexities of dual-use technologies like the Titan III launch vehicle.

At the time of MOL's cancellation, the program employed 192 service and 100 civilian personnel. The abrupt termination resulted in the reassignment of 80 percent of the service members and the transfer of civilian staff to the Space and Missile Systems Organization (SAMSO). Tragically, the program also lost several of its astronauts: Lieutenant John L. Finley had returned to the U.S. Navy in April 1968, and Major Michael J. Adams had perished in an F-104 crash at Edwards Air Force Base in July 1966. Major Robert H. Lawrence Jr., the first African-American astronaut, died in an F-104 crash on December 8, 1967. The remaining fourteen service personnel, with the exception of Lieutenant Colonel Robert T. Herres, sought transfers to NASA. NASA's Director of Flight Crew Operations, Deke Slayton, believed there was no immediate need for additional astronauts, while NASA Deputy Administrator George Mueller sought to maintain positive relations with the USAF. Ultimately, seven MOL pilots were absorbed into NASA's Astronaut Group 7 and later participated in Space Shuttle missions, beginning with Robert L. Crippen on STS-1. Others, like Albert H. Crews Jr., continued their careers as test pilots within NASA until 1994. The exposure to highly classified information restricted those who did not transfer to NASA from engaging in combat roles for three years to prevent the risk of capture, adversely affecting their military careers and leading some to leave the service altogether.

The technological advancements and infrastructural investments of the MOL program left an indelible mark on subsequent space endeavors. The Titan IIIM booster, although never flown, became a cornerstone of the military satellite program. Its successors, the Titan IIID and Titan IV, demonstrated its enduring legacy. The design of the Gemini B spacesuits influenced NASA's own suit designs, and MOL's waste management systems were integrated into the Skylab program. Additionally, the Space Shuttle's Solid Rocket Boosters incorporated materials, processes, and design elements initially developed for MOL's UA1207 solid rocket boosters. The MOL computer center's state-of-the-art systems were pivotal for managing complex space missions, and technologies developed for MOL found their way into NASA's Earth Science initiatives.

Furthermore, advancements in high-resolution optics for the MOL program contributed to the development of the Multiple Mirror Telescope in Arizona, one of the world's largest optical telescopes at the time of its dedication. The declassification of over eight hundred files and photos related to the MOL program in July 2015, coupled with oral histories and publications like Courtney V.K. Homer's Spies in Space (2019), have provided valuable insights into the program's operations and legacy.

The physical remnants of the MOL program serve as tangible reminders of its ambitious reach. The Gemini B spacecraft used in the sole MOL test flight is displayed at the Air Force Space and Missile Museum at Cape Canaveral Air Force Station, while another Gemini B spacecraft, modified for ground-based testing, is housed at the National Museum of the United States Air Force in Dayton, Ohio. Additionally, two MOL MH-7 training spacesuits were discovered in a locked room at the Cape Canaveral Launch Complex 5 museum in 2005, and one was donated by Robert L. Crippen to the National Air and Space Museum in 2017.

The infrastructure initially built for MOL also continued to influence later space programs. Space Launch Complex 6 at Vandenberg Air Force Base, nearly complete at the time of MOL's cancellation, was eventually repurposed for NASA's Space Shuttle program. Although the Space Shuttle Challenger disaster in January 1986 ended plans for polar orbit launches from SLC-6, the complex found new life with Delta IV launches from 2006 to 2022 and is presently being converted by SpaceX to support Falcon 9 and Falcon Heavy launches starting in 2025.

In summary, the Manned Orbiting Laboratory program, despite its premature termination, left a profound legacy on both military and civilian space endeavors. Its ambitious vision of a manned space station for reconnaissance missions drove significant technological advancements and fostered collaborations that would shape the future of space exploration. The lessons learned from MOL's development, including the challenges of balancing human and automated reconnaissance, the complexities of managing large-scale, semi-secret programs, and the importance of adaptable technological frameworks, continue to inform contemporary space initiatives. While MOL itself never achieved operational status, its influence persists, underscoring the intricate interplay between ambition, innovation, and pragmatism that has long characterized humanity's quest to explore and utilize the final frontier.

Chapter 11 - Anti-Satellite Weapons

United States

The US Space Force's stance on anti-satellite (ASAT) weapons is shaped by the growing threats to U.S. space assets, the need to preserve space as a safe and operational domain, and the recognition that space is increasingly becoming a contested environment. While the US Space Force is not publicly committed to developing and deploying ASAT weapons as a first-strike capability, it acknowledges the importance of defending against the increasing number of counter-space threats posed by nations such as Russia and China. This defensive posture reflects a broader commitment to maintaining space security, ensuring freedom of operation for U.S. assets, and deterring potential adversaries from hostile actions in space.

ASAT weapons, which are designed to disable or destroy satellites, are a central concern for the Space Force as space systems are vital to national security. The U.S. military and civilian and commercial sectors rely heavily on satellites for critical functions such as communications, navigation, intelligence gathering, and missile detection. The growing threat posed by the development of ASAT capabilities by other nations—particularly China and Russia—has raised alarms about the potential for space to become a battlefield in future conflicts. Both nations have demonstrated their ability to conduct ASAT operations, including direct-ascent missiles capable of destroying satellites and co-orbital systems that can interfere with or disable space assets.

In response to these growing threats, the US Space Force focuses on enhancing the resilience and protection of U.S. space infrastructure. This includes developing capabilities that ensure the survivability of satellites and other space-based assets in the face of adversarial actions. The Space Force's primary stance on ASAT weapons is one of deterrence and defense, seeking to dissuade potential aggressors from targeting U.S. satellites by maintaining the ability to protect and defend these assets.

While the US has not openly deployed or tested ASAT weapons in recent years, it maintains the capability to respond to hostile actions in space if necessary. The Space Force's emphasis on space resilience involves developing systems that can rapidly replace or recover from attacks on satellites. These strategies include the use of smaller, more maneuverable satellites, as well as the exploration of on-orbit servicing and repair capabilities, which could allow damaged satellites to be fixed or repositioned in the event of a hostile strike.

Furthermore, the US Space Force advocates for the responsible use of space and supports efforts to prevent its weaponization to the extent possible. This position aligns with the broader U.S. government policy, which encourages the peaceful use of space and adherence to international norms and treaties, such as the Outer Space Treaty. However, the US has made it clear that it reserves the right to defend its space assets and respond to any threat, including ASAT attacks, in a manner consistent with its national defense and strategic interests.

The Space Force is also deeply concerned about the consequences of ASAT weapons that create space debris. Kinetic ASAT tests, such as those conducted by China in 2007 and Russia in 2021, which resulted in large clouds of space debris, are viewed as particularly dangerous. Debris poses a long-term threat to all spacefaring nations, as it can indiscriminately damage satellites and spacecraft in orbit. As such, the Space Force and the broader U.S. defense community are vocal in opposing irresponsible ASAT tests that increase the risk of space becoming an unusable and dangerous domain.

The US Space Force's efforts to counter ASAT threats also include working closely with international allies and commercial partners. By strengthening alliances with countries like the United Kingdom, Japan, and Australia, and by working with private space companies, the U.S. aims to build a coalition that supports the shared goal of maintaining the security and safety of space. This collective approach increases resilience by promoting the exchange of information and technologies, coordinating space traffic management, and enhancing global space situational awareness (SSA).

Anti-satellite weapons (ASATs) have long stood as pivotal instruments in the arsenal of modern military strategy, designed to incapacitate or destroy satellites for both strategic and tactical objectives. These sophisticated space weapons, though never deployed in active warfare, have been demonstrated by a select group of nations—China, India, Russia, and the United States—each showcasing their capability by successfully targeting and eliminating their own satellites. Beyond their military utility, ASATs have also been employed to deorbit decommissioned satellites, highlighting their dual role in both defense and space management.

The strategic importance of ASATs is multifaceted. They serve as essential defensive measures against adversaries' space-based and nuclear weapon systems, effectively acting as force multipliers that can enhance the impact of a nuclear first strike. Moreover, ASATs function as countermeasures against anti-ballistic missile defenses, providing an asymmetric advantage against technologically superior opponents. In the realm of counter-value warfare, they target critical space assets, thereby undermining an adversary's operational capabilities in space.

However, the deployment and use of ASATs carry significant risks, particularly the generation of space debris. Each interception or destruction event produces fragments that can collide with other satellites, initiating a chain reaction known as the Kessler syndrome. This scenario envisions a densely cluttered space environment where debris from previous ASAT actions perpetually threatens the integrity and functionality of both existing and future satellite systems, potentially rendering low Earth orbit hazardous for all spacefaring activities.

The development of ASAT technology has evolved through various technological and strategic pathways, with the United States and the Soviet Union initially leading efforts in the late 1950s. The United States Air Force embarked on a series of advanced strategic missile projects under the designation Weapon System WS-199A. Among these initiatives was Martin's Bold Orion air-launched ballistic missile (ALBM) for the B-47 Stratojet, leveraging the rocket motor from the Sergeant missile. Between May 1958 and October 1959, twelve test launches were conducted. Although these early attempts met with limited success, leading to the termination of the ALBM project, the system was subsequently adapted by incorporating an Altair upper stage. This modification extended its range to 1,770 kilometers and repurposed it as an anti-satellite weapon. In a notable test, the modified Bold Orion successfully approached the Explorer 6 satellite to within 6.4 kilometers, demonstrating the potential effectiveness of a nuclear weapon in neutralizing a satellite, though it also underscored the limitations of conventional warheads for such missions.

Parallel to the Bold Orion project, Lockheed's High Virgo initiative emerged under WS-199A. Initially designed as another ALBM for the B-58 Hustler, High Virgo was similarly adapted for the anti-satellite role. On September 22, 1959, High Virgo attempted to intercept Explorer 5. However, the mission faced setbacks when communications with the missile were lost shortly after launch, and subsequent recovery efforts failed to determine the success of the interception. With the advent of the GAM-87 Skybolt project, these early ASAT endeavors were largely abandoned, although smaller-scale projects continued intermittently into the early 1970s.

The 1960s marked a significant shift towards the utilization of high-altitude nuclear explosions for satellite destruction. Early tests, such as the Hardtack Teak detonation in 1958, revealed the formidable effects of electromagnetic pulses (EMP) on electronic equipment. The subsequent Starfish Prime test in 1962, involving a 1.4-megaton warhead, not only damaged three satellites but also disrupted power transmission and communications across the Pacific. These outcomes underscored the potential of nuclear explosions to incapacitate space assets while also highlighting the broader implications for global infrastructure. Further weapons effect testing under the DOMINIC I series culminated in the deployment of the nuclear-armed Nike Zeus missile for ASAT purposes. Codenamed Mudflap and designated DM-15S, this missile was stationed at the Kwajalein Atoll until 1966, when the program transitioned to the USAF Thor-based Program 437 ASAT, operational until 1975.

In addition to missile-based ASAT systems, research into directed-energy weapons introduced another dimension to anti-satellite capabilities. In 1968, Lawrence Livermore National Laboratory (LLNL) proposed a nuclear-explosion powered X-ray laser, an ambitious concept that sought to leverage high-energy physics for space warfare. This initiative expanded into the development of conventional lasers and masers, exploring innovative ideas such as satellites equipped with fixed lasers and deployable mirrors for precise targeting. Although the X-ray laser system was ultimately canceled in 1977, the pursuit of directed-energy technologies was later revived during the 1980s as part of the Strategic Defense Initiative (SDI), reflecting ongoing interest in advanced ASAT solutions.

ASAT initiatives remained relatively low in priority until 1982, when revelations of a successful Soviet ASAT program galvanized the United States into action. This led to a "crash program" that culminated in the development of the Vought ASM-135 ASAT missile, based on the AGM-69 SRAM with an Altair upper stage. Deployed on a modified F-15 Eagle, the ASM-135 featured enhanced guidance systems and a data link for mid-course updates, enabling precise targeting capabilities. The missile made its first launch in January 1984, and on September 13, 1985, it achieved its first and only successful interception. Launched from Edwards Air Force Base, the missile struck the Solwind P78-1 gamma ray spectroscopy satellite at an altitude of 555 kilometers. This operation demonstrated the practical viability of direct-ascent ASATs, although it also generated space debris that eventually deorbited by May 2004. Despite its success, the ASM-135 program was discontinued in 1988, reflecting shifting strategic priorities and technological challenges.

The turn of the millennium saw continued advancements and operational tests of ASAT capabilities. A notable instance occurred on February 21, 2008, during Operation Burnt Frost, when the United States Navy deployed a ship-fired RIM-161 Standard Missile 3 to intercept and destroy USA-193, an American reconnaissance satellite that had malfunctioned shortly after its December 2006 launch. The interception took place approximately 247 kilometers above the Pacific Ocean, resulting in the creation of 174 pieces of orbital debris. While most of this debris re-entered the Earth's atmosphere within months, a few fragments persisted in higher orbits, posing lingering threats to space operations. The primary motivation for the mission was the approximately 450 kilograms of toxic hydrazine fuel aboard USA-193, which posed potential environmental and health risks if remnants survived re-entry. The successful destruction of the satellite underscored the enduring relevance of ASAT capabilities while also highlighting the environmental concerns associated with their use.

In a significant policy shift, the United States ceased testing direct-ascent anti-satellite missiles in 2022, signaling a move away from active ASAT development. This decision reflected evolving strategic priorities, international considerations regarding the militarization of space, and the growing recognition of the need to preserve the sustainability of the space environment. As nations continue to navigate the complexities of space security, the legacy of ASAT programs remains a critical consideration in shaping the future governance and utilization of outer space.

Star Wars - the Strategic Defense Initiative

The US Space Force's stance on the Strategic Defense Initiative (SDI), commonly known as "Star Wars," is shaped by its historical significance as an early attempt to integrate space-based defense systems into the United States' national security strategy. Although the SDI program was eventually scaled back and never fully realized, it laid the groundwork for key concepts in space-based defense that continue to influence the US Space Force's mission today.

The Strategic Defense Initiative was announced by President Ronald Reagan in 1983 during the height of the Cold War. The program's primary goal was to develop a missile defense system capable of protecting the United States from a large-scale nuclear attack, particularly from the Soviet Union. SDI was groundbreaking because it focused on space-based systems that could intercept and destroy incoming ballistic missiles before they reached

U.S. soil. These systems included proposals for ground- and space-based lasers, particle beams, and kinetic energy weapons that would be deployed in orbit to detect and eliminate nuclear threats.

Though the SDI program was ambitious and technologically advanced for its time, it faced significant challenges. The technical feasibility of deploying space-based weapons, combined with the immense costs of developing such systems, ultimately led to the program being scaled back in the early 1990s after the end of the Cold War. SDI was restructured under the name Ballistic Missile Defense Organization (BMDO), and its focus shifted toward more practical, earth-based missile defense systems. However, SDI's conceptual innovations influenced future missile defense programs and space-based military strategy.

The US Space Force, established in 2019, does not directly advocate for a revival of SDI's original vision of space-based missile interceptors. However, many of the principles that emerged from SDI—such as the importance of space for missile detection, tracking, and early warning—are central to the Space Force's current mission. Space is viewed as a critical domain for maintaining U.S. missile defense capabilities, especially with the advancements made in missile technologies by potential adversaries like China and Russia.

One key area where the legacy of SDI is evident in the Space Force's operations is in the domain of missile warning and tracking. The US Space Force operates a global network of satellites that provide early warning of ballistic missile launches. These satellites are equipped with infrared sensors that detect the heat signature of missile launches, allowing for real-time tracking and the ability to trigger defensive countermeasures. This type of early warning system is an evolution of some of the concepts envisioned during SDI, even though the actual deployment of space-based interceptors did not materialize.

While the Space Force does not pursue space-based weapons similar to those envisioned by SDI, it ensures that space remains a secure and operational domain for missile defense. This includes developing more advanced satellites and sensors for tracking missile threats and working closely with other branches of the U.S. military to integrate space assets into the broader missile defense architecture. These efforts reflect a more practical and technologically feasible approach to space-based defense compared to the ambitious scope of SDI.

Additionally, the US Space Force, like the broader Department of Defense, recognizes the importance of maintaining space superiority and preventing adversaries from deploying their own missile defense or offensive systems in space. With the increasing militarization of space, the Space Force's mandate includes protecting U.S. space assets from potential threats and ensuring that any missile defense systems operating in or through space are secure from interference or destruction by adversaries.

While the US Space Force does not directly pursue the goals of the Strategic Defense Initiative in its original form, the legacy of SDI's focus on space-based defense continues to shape the Space Force's approach to missile defense and space security. The Space Force emphasizes space as a key domain for early missile warning, tracking, and ensuring the security of U.S. space assets, reflecting the enduring influence of SDI's conceptual vision for defending against missile threats from space.

The dawn of the 1980s marked a transformative period in the geopolitical landscape of space warfare, largely influenced by the advent of the Strategic Defense Initiative (SDI). Proposed by President Ronald Reagan in 1983, SDI was primarily envisioned as a comprehensive missile defense system aimed at protecting the United States from nuclear ballistic missile attacks. However, the technological innovations and strategic imperatives underpinning SDI had profound implications beyond its original scope, notably invigorating anti-satellite weapon (ASAT) programs in both the United States and the Soviet Union.

SDI acted as a catalyst, providing significant impetus to existing ASAT initiatives and fostering the development of new technologies that could serve dual purposes in both missile defense and space-based offensive operations. The interconnection between ASAT and anti-ballistic missile (ABM) systems became increasingly apparent, leading to a symbiotic relationship where advancements in one domain could enhance capabilities in the other.

This convergence prompted both superpowers to reallocate resources and prioritize the integration of ASAT technologies within their broader strategic frameworks.

In the United States, the initial vision for SDI incorporated leveraging the already-developed Missile Homing Vehicle (MHV) as the cornerstone for a space-based constellation consisting of approximately forty platforms. These platforms were intended to deploy up to 1,500 kinetic interceptors, designed to engage and neutralize incoming ballistic missiles. By 1988, the scope of the US project had expanded into a more elaborate four-stage development plan. The first stage, known as Brilliant Pebbles, envisioned a satellite constellation comprising 4,600 kinetic interceptors, each weighing around 45 kilograms, stationed in low Earth orbit. These interceptors would work in tandem with sophisticated tracking systems to detect and engage ballistic threats with unprecedented precision.

Subsequent stages of the SDI program aimed to deploy larger and more capable platforms, incorporating advanced technologies such as laser and charged particle beam weapons. Projects like the Mid-Infrared Advanced Chemical Laser (MIRACL) were integral to these developments, representing the cutting edge of directed-energy weapon research. The ambitious timeline set for the first stage anticipated completion by the year 2000, with projected costs nearing $125 billion. This monumental investment underscored the United States' commitment to achieving a technologically superior and strategically advantageous position in space.

Meanwhile, the Soviet Union was compelled to respond to the burgeoning US advancements in space-based defense mechanisms. Recognizing the strategic threat posed by SDI, Soviet authorities initiated substantial investments in their own ASAT research and development programs. The 12th Five Year Plan saw a concerted effort to consolidate various research endeavors under the auspices of the State Committee for Defense Technology (GUKOS), aligning Soviet ASAT capabilities with the projected US deployment timelines. This period of intensified focus culminated in the experimental launch of the Polyus spacecraft in the late 1980s. Designed as an orbital laser platform, Polyus aimed to demonstrate the feasibility of directed-energy ASATs capable of disabling or destroying enemy satellites. However, the mission ended in failure when the spacecraft failed to achieve orbit, reflecting the significant technical challenges inherent in such ambitious projects.

As the Cold War neared its conclusion, both the United States and the Soviet Union began to reassess and subsequently reduce their expenditures on ASAT and SDI-related projects starting in 1989. The dissolution of the Soviet Union in 1991 led to the Russian Federation unilaterally discontinuing all SDI research by 1992, driven by economic constraints and shifting strategic priorities. Despite this decline, reports emerged indicating a resurgence of interest in ASAT technologies under President Vladimir Putin's administration. These efforts were purportedly motivated by the need to counteract renewed US strategic defense initiatives that emerged following the dissolution of the Anti-Ballistic Missile Treaty. However, the precise status and funding mechanisms of these contemporary Russian ASAT programs remain shrouded in secrecy, often intertwined with classified projects managed by the National Reconnaissance Office.

In parallel, the United States continued to explore foundational technologies that could underpin future space-based ASAT capabilities. Programs such as the Experimental Spacecraft System (USA-165), the Near Field Infrared Experiment (NFIRE), and the Space-Based Interceptor (SBI) represent ongoing efforts to develop advanced systems capable of asserting dominance in space. These initiatives focus on enhancing detection, tracking, and interception technologies, ensuring that the United States maintains a strategic edge in the increasingly contested space domain.

The United States' development of anti-satellite (ASAT) systems dates back to the early 1960s, reflecting the growing recognition of space as a critical domain for national security and military strategy. During this period, the United States embarked on several programs aimed at developing technologies capable of disabling or destroying enemy satellites, thereby securing its own space assets and maintaining superiority in the increasingly contested extraterrestrial arena.

In the early 1960s, the United States Air Force initiated the development of ground-based ASAT systems. These systems primarily utilized Thor missiles equipped with clear warheads, launched from strategic locations such as Johnston Island. Concurrently, the Army focused on deploying Nike-Zeus missiles from Kwajalein Atoll in the Pacific. The Army's ASAT tests began in May 1963, marking the nation's first significant foray into space-based weaponry. However, despite these efforts, the Army's system was deactivated by 1964, indicating early challenges in achieving reliable and effective ASAT capabilities.

Meanwhile, the Air Force continued its ASAT endeavors, conducting tests of its missile systems starting in May 1964. These efforts maintained partial operational status until 1975, demonstrating a sustained commitment to refining kinetic ASAT technologies. The Air Force focused on direct-fire systems, where missiles were launched from aircraft such as the F-15 equipped for ASAT roles. This integration of ASAT capabilities into existing military platforms highlighted the strategic importance placed on space dominance.

As the Cold War intensified, the Department of Defense (DOD) expanded its ASAT programs to incorporate more advanced and versatile technologies. One notable initiative involved developing manned vehicles (MHVs) to launch ASAT weapons from two-stage rockets. These systems were designed to employ short-range attack mechanisms (SRAK) and long-range systems, enhancing the United States' ability to engage and neutralize adversary satellites from various vectors.

In the fiscal year 1983 (FY83), Congress authorized substantial funding for advancing space-based laser technologies, recognizing their potential as non-kinetic ASAT weapons. The House of Representatives requested $148.8 million for ASAT development, while the Department of Defense appropriated $218 million to support these initiatives. These funds were allocated towards developing chemical lasers and short-wavelength laser technologies, which were expected to provide precise and effective means of disabling satellites without the need for direct physical destruction.

Through its appropriations process, the Senate supported the continuation and expansion of these research programs. Although specific proposals for space-based laser weapons systems, such as the Talon Gold program, were initially denied, adjustments and reallocations of funds ensured ongoing progress in ASAT technology. By September 1982, the FY83 DOD appropriation bill included provisions for furthering space-based laser research, demonstrating bipartisan support for maintaining technological superiority in space warfare.

Throughout the 1980s and beyond, the United States continued to invest in and develop a diverse array of ASAT systems. Programs like SEESAW and Chair Heritage, initiated in the late 1950s and early 1970s respectively, laid the groundwork for subsequent advancements in particle beam research and laser technology. These efforts culminated in establishing programs such as SIPA POU White Horse, which focused on refining laser capabilities for space-based applications.

By the late 20th century, the United States had developed a multifaceted ASAT arsenal encompassing both kinetic and energy-based systems. Space-based lasers, high-powered directed-energy weapons, and advanced missile interceptors formed the core of these capabilities, enabling the United States to project power and defend its interests in space effectively. The integration of these technologies into military operations underscored the strategic imperative of maintaining space superiority amidst evolving global threats.

The interplay between SDI and ASAT programs exemplifies the intricate relationship between defensive and offensive space technologies during the latter half of the 20th century. The relentless pursuit of strategic superiority in space not only fueled an arms race that extended beyond Earth's atmosphere but also highlighted the profound implications of space weaponization. As nations continue to navigate the complexities of space security, the legacy of SDI and the corresponding ASAT advancements serve as a testament to the enduring significance of space as a critical frontier in global military strategy.

The survivability of critical U.S. military satellite systems has become an increasingly pressing concern in the absence of an anti-satellite (ASAT) limitation agreement. As tensions escalate, ensuring that essential satellite operations can withstand potential ASAT attacks has become a focal point of strategic defense initiatives.

Several methods have been developed to enhance the survivability of these vital space assets. A primary vulnerability lies in the power systems of military satellites, most of which rely on solar panels. These panels are susceptible to shrapnel attacks, which could severely impair or disable the satellite's functionality. To mitigate this risk, the use of radioisotope thermoelectric generators (RTGs) is being explored as an alternative to solar cells. Unlike solar panels, RTGs can be housed within the spacecraft, reducing their exposure to external threats and thereby enhancing the satellite's resilience against kinetic attacks.

Precision targeting systems are another critical component in safeguarding military satellites. For an ASAT weapon to effectively neutralize a satellite, it must possess a highly accurate targeting mechanism capable of striking the satellite directly, rather than merely causing nearby explosions. This level of precision ensures that the satellite itself is impacted, minimizing collateral damage and maintaining the integrity of surrounding space infrastructure.

In addition to shielding and targeting improvements, maneuvering capabilities are being integrated into critical satellites to increase their chances of evading potential interceptors. By equipping satellites with the ability to alter their orbits, the likelihood of successfully avoiding an ASAT attack is significantly enhanced. Adequate warning systems are essential for this strategy, providing sufficient time for satellites to execute evasive maneuvers. To support this, the U.S. Air Force is investing in advanced space surveillance systems aimed at better monitoring satellite activities. One such system, the Ground-Based Electro-Optical Deep Space Surveillance (GEODSS), is being procured to become operational, offering enhanced detection and tracking of satellites in various orbits.

Ground-based radar systems, already in use, are also undergoing upgrades to bolster their effectiveness in identifying and tracking potential threats. Furthermore, the development of a space-based surveillance network is underway, which would comprise satellites positioned in both geosynchronous and low Earth orbits. This network aims to provide comprehensive monitoring capabilities, enabling more precise and timely detection of ASAT activities.

Another innovative approach to enhancing satellite survivability involves the deployment of separate, non-detectable satellites as backups or decoys. These satellites would be placed in very high orbits, possibly reaching altitudes of up to 115,000 kilometers, and designed to remain undetected by conventional radar systems. By maintaining these satellites in powered-down modes, they would evade detection by infrared sensors until they are needed to replace or protect primary satellite systems in the event of an attack. Decoy satellites can also serve to confuse and misdirect ASAT efforts, further safeguarding critical communication and reconnaissance functions.

In addition to these measures, efforts are being made to harden satellites against various forms of radiation and to equip them with defensive systems. The construction of space-based laser defense systems is also being considered as a means to protect satellites from incoming ASAT threats. These defensive technologies aim to neutralize or destroy attacking ASAT weapons before they can reach their intended targets, providing an additional layer of security for essential space assets.

Overall, the United States is employing a multifaceted strategy to enhance the survivability of its critical military satellite systems. By improving power source resilience, integrating maneuvering capabilities, advancing surveillance and tracking technologies, deploying decoy satellites, and developing defensive measures, the U.S. aims to ensure that its space-based infrastructure remains operational and secure in the face of evolving ASAT threats. These comprehensive efforts underscore the strategic importance of space as a contested domain and reflect the ongoing commitment to maintaining space superiority in an increasingly complex global landscape.

Soviet Union

The Soviet Union, recognizing the strategic imperative of controlling space, embarked on the development of anti-satellite weapons (ASATs) in response to the burgeoning threats posed by both bombardment satellites and ballistic missiles. This initiative was driven by the dual necessity of defending Soviet space assets and countering potential adversaries' capabilities in space warfare.

The genesis of the Soviet ASAT program can be traced back to the mid-1950s. While some accounts credit Sergei Korolev and his design bureau, OKB-1, with initiating early conceptual work on ASATs around 1956, others attribute the foundational efforts to Vladimir Chelomei's OKB-52 starting in 1959. What remains indisputable is that in April 1960, a pivotal meeting convened by Premier Nikita Khrushchev at his summer residence in Crimea laid the groundwork for the Soviet Union's foray into space-based defensive weaponry. During this assembly, Chelomei presented his comprehensive rocket and spacecraft program, securing approval to develop the UR-200 rocket. This versatile launch vehicle was designated to serve multiple roles, including as the launcher for the nascent anti-satellite project.

By March 1961, the decision to advance the ASAT initiative was formalized under the program name Istrebitel Sputnikov (IS), translating to "destroyer of satellites." The IS system was ingeniously designed as a co-orbital weapon, meaning it would approach a target satellite gradually, culminating in the detonation of a shrapnel-laden warhead in close proximity to neutralize the satellite. The operational procedure involved launching the missile when a target satellite's ground track ascended above the launch site. Once identified, the interceptor missile would be propelled into an orbit that brought it within striking distance of the intended satellite, typically within 90 to 200 minutes or approximately one to two orbits. Guided by an onboard radar system, the 1,400-kilogram interceptor was capable of engaging targets within a kilometer's range, making it a formidable tool in the Soviet space defense arsenal.

However, the development of the IS system was not without challenges. Delays in the UR-200 missile program necessitated an alternative approach, prompting Chelomei to seek the use of R-7 rockets for prototype testing. This led to two critical test launches on November 1, 1963, and April 12, 1964. Despite these efforts, the program faced a significant setback when Khrushchev, later in 1964, opted to cancel the UR-200 in favor of the more advanced R-36 rocket. This strategic pivot forced the IS program to adapt, transitioning to the R-36-derived Tsyklon-2 launcher. Nonetheless, delays persisted, leading to the development of a simplified version known as the 2A. This iteration successfully launched its first IS test on October 27, 1967, followed by a second test on April 28, 1968. These tests targeted the specially designed DS-P1-M spacecraft, with the IS warhead's shrapnel demonstrating its destructive capability by recording multiple hits. In total, 23 launches constituted the IS test series, culminating in the system's declaration of operational readiness in February 1973.

The Soviet ASAT program achieved a landmark victory in February 1970 with the world's first successful intercept of a satellite. This initial success was further solidified by a subsequent test that achieved 32 direct hits, each capable of penetrating 100 millimeters of armor, underscoring the system's efficacy. However, the momentum of the program experienced a resurgence in 1976, driven by fears within the Soviet leadership about the potential militarization of space by the United States, particularly with the advent of the Space Shuttle. Influenced by claims that the Shuttle could function as a single-orbit weapon capable of evading existing anti-ballistic missile defenses and striking strategic targets such as Moscow, Soviet Premier Leonid Brezhnev ordered the expansion of the IS program. This expansion aimed to enhance the system's capabilities to engage targets at higher altitudes, leading to its operational declaration in this new configuration on July 1, 1979.

Despite these advancements, the Soviet ASAT program faced its ultimate decline in 1983 under the leadership of Yuri Andropov, who ordered the cessation of all IS testing. Subsequent attempts to revive the program were unsuccessful, coinciding ironically with the United States initiating its own ASAT testing efforts in response to the Soviet advancements.

In parallel with the IS system, the Soviet Union explored alternative ASAT technologies in the early 1980s. One notable development was the creation of the 30P6 "Kontakt" system, which mirrored the United States' air-launched ASAT capabilities. Utilizing modified MiG-31D 'Foxhound' interceptors as launch platforms, at least six units of the 79M6 missile were completed. Additionally, the Soviets experimented with arming their Almaz space stations with Rikhter R-23 aircraft auto-cannons, aiming to provide space-based defensive firepower. Another ambitious project was the 11F19DM Skif-DM/Polyus, an orbital megawatt laser system intended to serve as a directed-energy ASAT. However, this project failed to achieve its objectives upon its launch in 1987.

In 1987, under the leadership of Mikhail Gorbachev, the Soviet Union showcased the "Naryad" (Sentry) ASAT system, also known by its designation 14F11. This system was launched using UR-100N rockets from the Baikonur Cosmodrome, representing the latest iteration of Soviet efforts to maintain parity with American ASAT capabilities. Despite these developments, the dissolution of the Soviet Union in 1991 brought an end to these advanced ASAT projects, as the newly formed Russian Federation reprioritized its military and space initiatives amidst changing geopolitical landscapes.

Throughout its tenure, the Soviet ASAT program was characterized by a relentless pursuit of technological innovation and strategic dominance in space. The legacy of these efforts not only influenced subsequent Russian space defense strategies but also underscored the profound implications of space weaponization, a theme that continues to resonate in contemporary discussions about the militarization of space.

Following the dissolution of the Soviet Union in 1991, Russia experienced significant reductions in defense expenditures, leading to the temporary suspension of several advanced military projects, including the MiG-31D interceptor. This strategic pause reflected the broader economic and geopolitical shifts that the newly formed Russian Federation was navigating during the tumultuous post-Cold War era. However, the resurgence of global tensions and the evolving nature of space as a contested domain prompted a renewed interest in anti-satellite (ASAT) capabilities.

In August 2009, Alexander Zelin, a prominent figure within the Russian defense establishment, announced the resumption of the MiG-31D project. This move signaled Russia's commitment to reestablishing its presence in space-based defense technologies. Concurrently, the development of the Sokol Eshelon emerged as a critical component of Russia's ASAT strategy. The Sokol Eshelon, a prototype laser system mounted on an A-60 aircraft, was reported to restart development in 2012. This system aimed to harness directed-energy technologies to disable or destroy enemy satellites, representing a significant advancement in Russia's space weaponry.

The following years saw a series of strategic ASAT tests that underscored Russia's revitalized focus on space dominance. In December 2016, three additional ASAT launches were reportedly conducted, demonstrating the operational readiness and expansion of Russia's ASAT arsenal. These tests continued with launches on March 26, 2018, and December 23, 2018, the latter two executed from transporter erector launchers (TELs), highlighting Russia's capability to deploy ASAT missiles from mobile platforms. September 2018 marked a notable milestone when a new type of ASAT missile was observed being carried by a MiG-31 aircraft, indicating advancements in both missile technology and deployment strategies.

The early 2020s further illustrated Russia's persistent advancements in ASAT technologies. On April 15, 2020, U.S. officials reported that Russia had conducted a direct ascent ASAT missile test capable of targeting spacecraft or satellites in low Earth orbit. This test was followed by another launch on December 16, 2020, reinforcing Russia's commitment to maintaining and enhancing its space-based offensive capabilities. The culmination of these efforts was evident in November 2021, when Russia successfully destroyed Kosmos 1408, an ASAT missile test that resulted in a significant debris field. This debris posed a tangible threat to the International Space Station

(ISS), underscoring the broader implications of ASAT activities on space sustainability and the safety of orbital operations.

In the lead-up to 2024, U.S. intelligence sources indicated that Russia was developing an ASAT weapon incorporating nuclear technology. While it remained unclear whether this system employed a nuclear warhead or utilized nuclear power to enhance its destructive capabilities, the development highlighted the escalating sophistication and potential lethality of Russia's ASAT arsenal. This progression not only demonstrated Russia's ongoing commitment to space militarization but also emphasized the critical need for international dialogue and regulatory frameworks to mitigate the risks associated with the weaponization of space.

Throughout these developments, Russia's ASAT program has evolved in response to shifting geopolitical dynamics and technological advancements. From the initial suspension of projects in the post-Soviet era to the aggressive reactivation and enhancement of ASAT capabilities in the 21st century, Russia has consistently sought to secure its strategic interests in the increasingly contested domain of space. These efforts reflect a broader global trend towards the militarization of space, where the control and protection of space assets have become paramount to national security and technological supremacy.

China

Parallel to the advancements made by the United States and Russia, the People's Republic of China emerged as a formidable player in the development of anti-satellite (ASAT) technologies, further intensifying the global Space Race. China's commitment to establishing dominance in space became evident through a series of deliberate and strategic ASAT tests, underscoring its ambition to safeguard its space assets while simultaneously deterring potential adversaries.

A pivotal moment in China's ASAT program occurred on January 11, 2007, when the nation successfully destroyed its own defunct weather satellite, Fengyun-1C (FY-1C). This operation marked China's first demonstrated capability to neutralize a satellite using kinetic energy, employing an SC-19 ASAT missile equipped with a kinetic kill vehicle analogous to the American Exoatmospheric Kill Vehicle. FY-1C, launched in 1999 as the fourth satellite in the Fengyun series, orbited Earth in a polar orbit at an altitude of approximately 865 kilometers and had a mass of around 750 kilograms. The missile was deployed from a mobile Transporter-Erector-Launcher (TEL) vehicle stationed in Xichang, coordinates 28.247°N 102.025°E. The SC-19 missile achieved its objective by colliding head-on with the satellite at an extremely high relative velocity, resulting in the fragmentation of FY-1C.

The destruction of FY-1C generated a significant amount of space debris, with over 40,000 new fragments measuring larger than one centimeter dispersing into orbit. This incident not only demonstrated China's ASAT capabilities but also highlighted the potential for such actions to exacerbate the existing problem of space debris, posing risks to other satellites and the broader space environment.

Subsequent tests of the SC-19 system further evidenced China's commitment to developing ASAT technologies. Reports indicate that similar ASAT missions were conducted in 2005, 2006, 2010, and 2013, each reinforcing China's proficiency in space-based weaponry. These repeated demonstrations served both as showcases of technological prowess and as strategic deterrents, signaling to other nations the robustness of China's space defense mechanisms.

In May 2013, the Chinese government publicly announced the launch of a suborbital rocket intended to carry a scientific payload for studying the upper ionosphere. However, intelligence assessments from the United States suggested that this mission was, in fact, a covert test of a new ground-based ASAT system. An open-source analysis conducted by the Secure World Foundation, which utilized commercial satellite imagery, supported this interpretation, indicating that the rocket launch could potentially threaten US satellites in geostationary Earth orbit. This duality between scientific advancement and military application exemplifies the complex interplay between civilian and defense-oriented space technologies.

China continued to refine its ASAT capabilities, conducting another significant test on February 5, 2018. On this date, China launched an exoatmospheric ballistic missile, the Dong Neng-3, purportedly as part of a defensive initiative. State media reported that the test achieved its intended objectives without any offensive implications. However, the strategic implications of such tests remain profound, as they demonstrate China's ability to deploy high-velocity interceptors capable of targeting satellites beyond Earth's atmosphere.

These developments underscore China's strategic emphasis on securing its interests in space through the development of advanced ASAT systems. By enhancing its ability to disable or destroy adversarial satellites, China not only protects its own space assets but also gains leverage in potential conflicts that extend into the space domain. The cumulative effect of these tests contributes to the evolving dynamics of space security, where the ability to control and defend space assets becomes increasingly critical to national security and technological supremacy.

The trajectory of China's ASAT program reflects a broader trend of militarization in space, where major powers invest in space-based weaponry to assert dominance and ensure the protection of their extraterrestrial interests. As China continues to advance its ASAT technologies, the international community faces growing concerns about the sustainability and safety of space operations. The proliferation of ASAT capabilities among global powers elevates the stakes of space competition, necessitating robust dialogue and regulatory measures to prevent the escalation of space weaponization and to preserve the orbital environment for peaceful and cooperative use.

China's pursuit of anti-satellite (ASAT) capabilities began in the early 1960s, marking a significant chapter in the broader narrative of the Space Race. In 1964, the People's Republic of China initiated what would become a long-term strategic effort to develop technologies capable of disabling or destroying enemy satellites. This endeavor was initially conducted under Program 640, a government initiative primarily focused on anti-ballistic missile development and surface-to-air missile (SAM) sites. However, by 1970, Program 640 had expanded its scope to include the nascent ASAT program, reflecting China's growing recognition of space as a critical domain for military advantage.

The progress of China's early ASAT efforts was substantially hindered by the tumultuous period of the Cultural Revolution. During this time, many of the nation's leading scientists and engineers, who were integral to advancing ASAT technologies, were purged by the younger generation that dominated the political landscape. This internal strife not only slowed technological development but also led to a concentration of knowledge, with a single prominent scientist authoring the majority of scholarly articles related to ASAT systems in the 1970s. The disruption caused by the Cultural Revolution left Program 640 struggling to maintain momentum in its ambitious military objectives.

By 1980, the limitations imposed by the Cultural Revolution had taken their toll, leading to the abandonment of Program 640. It was not until six years later, in 1986, that the Chinese government restructured its technological research and development efforts under the banner of Program 863. This new initiative marked a renewed commitment to advancing China's military and space capabilities. Program 863 became closely associated with the General Armaments Department's Project 921 and the State Administration for Science, Technology and Industry for National Defense (SASTIND), formerly known as the Commission for Science, Technology and Industry for National Defense. These organizations played pivotal roles in coordinating and advancing China's counterspace technologies.

Throughout the subsequent decades, China's ASAT program made significant strides, developing three distinct systems capable of targeting satellites: direct fire ASAT missiles, directed-energy weapons, and microsatellites. These systems underwent various tests, some of which were publicly acknowledged by the Chinese government, while others were reported by external observers as demonstrating ASAT capabilities. The evolution of these technologies underscored China's commitment to establishing a robust and versatile counterspace arsenal.

By the early 21st century, China's efforts had expanded to encompass a comprehensive array of counterspace capabilities. This included direct-ascent ASAT missiles, which are launched directly from the ground to intercept satellites; co-orbital ASAT systems, which involve deploying satellites equipped with the means to disable or destroy other satellites while in orbit; computer network operations aimed at disrupting satellite communications and control; ground-based satellite jammers that interfere with the functioning of space-based assets; and advanced directed-energy weapons capable of damaging or destroying satellites without traditional kinetic impacts.

The People's Liberation Army (PLA) took an active role in integrating these technologies into their operational framework. Military units were formed, and initial operational training commenced, focusing on the deployment and use of counterspace capabilities such as ground-launched ASAT missiles. This integration signified a strategic shift, positioning China's military to leverage space as a contested and critical theater of modern warfare.

In 2008, during the development of the JL-2 submarine-launched ballistic missile, reports emerged that China was exploring modifications to enable the missile to carry an ASAT warhead. This adaptation aimed to provide China with a sea-based ASAT capability, thereby diversifying its methods for targeting and neutralizing enemy satellites. Such advancements highlighted China's determination to enhance its strategic flexibility and ensure dominance in the increasingly contested space domain.

The culmination of these efforts was evident in 2016 when a report by the United States Congress raised alarms over China's development of space weapons designed to destroy American satellites. This report underscored the growing tensions and the potential for space to become a new frontier of military confrontation. China's sustained investment in ASAT technologies not only demonstrated its technical prowess but also its strategic intent to secure and expand its influence in space, positioning itself as a formidable player in the ongoing global competition for dominance in this final frontier.

China's commitment to developing a comprehensive suite of anti-satellite (ASAT) systems reflects its strategic emphasis on space as a critical domain for national security. Over the years, the People's Republic of China has pursued three primary categories of ASAT technologies: direct fire systems, directed-energy weapons, and microsatellites. Each of these systems represents a distinct approach to disabling or destroying enemy satellites, underscoring the multifaceted nature of China's counterspace capabilities.

Direct fire systems form the backbone of China's kinetic ASAT strategy. These systems involve the use of land- or vehicle-based missiles specifically designed to physically strike and incapacitate satellites. A notable demonstration of this capability occurred on January 11, 2007, when China successfully launched an SC-19 missile to destroy an aging Fengyun series weather satellite. This operation not only showcased China's ability to execute precise kinetic strikes from terrestrial platforms but also highlighted the potential vulnerabilities of space assets to such direct assaults. Although there have been no known tests of vehicle-based direct fire systems beyond this instance, reports indicate that China's emerging Jin-class submarines may soon possess the capability to deploy SC-19 missiles or similar types, thereby extending ASAT capabilities to sea-based platforms. This diversification of launch platforms enhances China's strategic flexibility, allowing it to threaten satellites from multiple vectors.

In parallel with kinetic approaches, China has been actively developing directed-energy weapons (DEWs) as part of its New Concept Weapons program. DEWs, which include high-powered lasers and microwave weapons, offer the advantage of disabling or damaging satellites without the need for physical destruction. Since 1995, China has invested in the development of high-powered laser systems capable of targeting spaceborne assets. One significant milestone was the 2006 test of such a laser against orbiting U.S. satellites, demonstrating the potential of DEWs to interfere with or impair the functionality of enemy satellites. According to the United States Department of Defense, Chinese defense research has proposed several DEW systems capable of both reversible dazzling of electro-optical sensors and the irreversible destruction of satellite components. Projections suggest that by the mid to late 2020s, China will be able to field more advanced DEW systems, posing significant threats to non-optical satellites and expanding the scope of its counterspace operations.

Microsatellites represent a more recent and technologically sophisticated addition to China's ASAT arsenal. Defined as objects orbiting the Earth with masses between 10 and 500 kilograms, microsatellites are inherently versatile and can be weaponized with relative ease compared to larger satellites. Their high relative velocities in orbit mean that even a minor collision can result in the catastrophic destruction of both the target and the attacking satellite. In 2001, Chinese media reported that the country was experimenting with parasitic microsatellites designed to latch onto and disable other satellites on command. While direct evidence of such parasite systems remains elusive, incidents like the 2008 close approach of China's BX-1 microsatellite to the International Space Station (ISS) have raised concerns about the potential for microsatellites to be used as kinetic kill vehicles. The BX-1's dangerously close passage to the ISS, which could have resulted in mutual destruction had a collision occurred, served as a stark demonstration of the risks posed by weaponized microsatellites and underscored China's growing capabilities in this domain.

The integration of these three ASAT systems into China's military strategy signifies a deliberate and calculated effort to secure dominance in space. By developing direct fire systems, directed-energy weapons, and microsatellites, China aims to create a layered and resilient counterspace capability that can effectively neutralize a wide range of adversary satellites. The People's Liberation Army (PLA) has actively incorporated these technologies into its operational framework, establishing specialized military units and conducting initial training to ensure readiness. The ongoing advancements in ASAT technologies not only enhance China's ability to protect its own space assets but also provide it with the tools to disrupt and challenge the space-based infrastructure of potential adversaries.

China's anti-satellite (ASAT) operations over the past two decades illustrate a determined and methodical approach to establishing dominance in space. These operations, acknowledged either by the Chinese government or reported by foreign entities, demonstrate the evolution and sophistication of China's counterspace capabilities.

In the mid-2000s, China's ASAT endeavors began to gain international attention. In 2005, the People's Republic of China made its first notable attempt to employ a direct fire ASAT weapon. On July 7 of that year, a missile was launched with the intent to intercept a satellite. However, the mission did not achieve its objective, as the missile failed to approach the satellite closely enough to effect a strike. Undeterred, China conducted a second attempt on February 6, 2006. This time, the missile approached the satellite sufficiently to raise suspicions among observers about whether the miss was intentional or indicative of emerging technical challenges. These early attempts highlighted both China's growing ambition and the complexities involved in developing reliable kinetic ASAT systems.

Simultaneously, in 2006, the United States reported that China had directed a high-powered laser at American observation satellites. While no significant damage was observed, the incident underscored China's exploration of directed-energy weapons (DEWs) as a non-kinetic means of disabling space assets. The laser's purpose may have been dual: serving as a ranging tool to precisely determine satellite orbits and as a potential weapon to blind or disrupt satellite sensors. This development marked China's entry into the realm of energy-based counterspace technologies, expanding the scope of its ASAT program beyond traditional missile-based approaches.

The most consequential ASAT operation occurred on January 11, 2007, when China conducted a successful kinetic kill test. Utilizing an SC-19 missile launched from the Xichang Satellite Launch Center, China targeted and destroyed the FY-1C weather satellite, part of the Fengyun series, which orbited at an altitude of 865 kilometers. The kinetic kill vehicle traveled at an impressive speed of 8 kilometers per second, engaging the satellite in a head-on collision that obliterated the aging spacecraft. Initially unconfirmed by the Chinese government, the test was first reported by Aviation Week & Space Technology magazine and later verified by the United States National

Security Council. On January 23, 2007, the Chinese Foreign Ministry officially acknowledged the test, asserting that notifications had been sent to the United States, Japan, and other countries in advance. This operation not only demonstrated China's capability to execute precise kinetic strikes from land-based platforms but also served as a stark reminder of the vulnerabilities inherent in space-based infrastructure.

In September 2008, China further showcased its ASAT capabilities through the deployment of the BX-1 microsatellite. Released during the Shenzhou-7 mission, the BX-1 maneuvered perilously close to the International Space Station (ISS), passing within 27 miles at a relative speed of 17,000 miles per hour. Although a collision was narrowly avoided, the incident highlighted China's proficiency in deploying microsatellites with potential ASAT functionalities. The BX-1's close approach raised alarms about the risks posed by such satellites, which are smaller, more maneuverable, and cost-effective compared to their larger counterparts. This event underscored the dual-use nature of microsatellites, which can be employed for both peaceful purposes and as kinetic kill vehicles capable of disabling or destroying other space assets.

China continued to refine its ASAT technologies in the following years. On January 11, 2010, exactly three years after the successful SC-19 test, China launched another SC-19 missile. This time, the target was a medium-range ballistic missile, the CSS-X-11, launched from the Shuangchengzi Space and Missile Center. Although the primary objective of this test remains unclear, it signified a possible shift towards repurposing the SC-19 missile for anti-ballistic missile (ABM) applications, while still maintaining its ASAT capabilities. This dual functionality reflects China's strategic intent to create versatile missile systems capable of addressing multiple defense and offensive requirements.

The advancement of China's ASAT program was further evidenced by the development and testing of the Dong Neng series of ASAT interceptors. On May 13, 2013, China conducted a test launch from the Xichang Satellite Launch Center, identified by the United States as the 'Kunpeng-7' mission. This launch was part of the Dong Neng-2 (DN-2) series, designed to intercept and neutralize enemy satellites through kinetic means. The following year, on July 23, 2014, China reported a successful land-based missile test of an ASAT weapon. However, the United States characterized this test as "non-destructive," suggesting that while the missile demonstrated ASAT potential, it did not result in the destruction of a satellite. These tests illustrate China's ongoing efforts to enhance the reliability, precision, and versatility of its ASAT systems, ensuring readiness for a variety of operational scenarios.

Throughout these operations, China's strategy has been characterized by incremental advancements and strategic diversification. From initial direct fire attempts and the deployment of kinetic kill vehicles to the exploration of directed-energy weapons and the utilization of microsatellites, China has systematically expanded its counterspace arsenal. Each successful test and every close call in space have contributed to a more robust and multifaceted ASAT program, positioning China as a formidable player in the global Space Race.

India

As the global landscape of space militarization expanded, India emerged as a significant contender in the development of anti-satellite (ASAT) technologies, marking its entry into an exclusive group of nations capable of deploying space-based weaponry. India's foray into ASAT capabilities was driven by a combination of strategic defense imperatives, technological advancements, and the desire to assert its sovereignty in the increasingly contested domain of space.

The initial public acknowledgment of India's ASAT ambitions surfaced during the 97th Indian Science Congress held in Thiruvananthapuram on January 11, 2007. In a televised press briefing, Rupesh, the Director General of the Defence Research and Development Organisation (DRDO), announced that India was actively developing the requisite technologies to integrate a weapon system capable of neutralizing enemy satellites in orbit. This declaration was further substantiated on February 10, 2010, when Dr. Vijay Kumar Saraswat, DRDO Director-General and Scientific Advisor to the Defence Minister, affirmed that India possessed "all the building blocks necessary" to construct an ASAT weapon capable of targeting hostile satellites in both low Earth and polar orbits.

By April 2012, the momentum behind India's ASAT program had gained substantial traction. V. K. Saraswat, then-chairman of DRDO, declared that India had developed the critical technologies required for an ASAT weapon, leveraging advancements from the Indian Ballistic Missile Defence Programme. This period also saw strategic discourse within India's defense community, with experts like Ajay Lele from the Institute for Defence Studies and Analyses advocating for the benefits of conducting ASAT tests. Lele posited that demonstrating ASAT capabilities would strengthen India's position in international negotiations, particularly if an international regime akin to the Nuclear Non-Proliferation Treaty (NPT) were to be established to control ASAT proliferation. He suggested that a test targeting a purpose-launched satellite in low orbit would be perceived as a responsible demonstration of technological prowess rather than an act of recklessness. The culmination of these efforts was the formal sanctioning of the ASAT program in 2017, signaling India's commitment to advancing its space defense capabilities.

The definitive milestone in India's ASAT journey was achieved on March 27, 2019, with the successful execution of Mission Shakti. This operation demonstrated India's ability to deploy an ASAT missile capable of intercepting and destroying a satellite in low Earth orbit (LEO). Launched at approximately 05:40 UTC from the Integrated Test Range (ITR) in Chandipur, Odisha, the interceptor missile successfully struck the test satellite, Microsat-R, at an altitude of 300 kilometers after a flight duration of 168 seconds. Developed by DRDO, Mission Shakti established India as the fourth nation with demonstrated ASAT capabilities, following the United States, Russia, and China.

The strategic rationale behind Mission Shakti was multifaceted. India emphasized that the capability was intended as a deterrent rather than a weapon directed against any specific nation. In the aftermath of the test, the Indian Ministry of External Affairs released a statement assuring the international community that the operation was conducted at a low altitude to minimize the generation of long-lived space debris. According to Jonathan McDowell, an astrophysicist at the Harvard–Smithsonian Center for Astrophysics, most of the debris from the test was expected to re-enter Earth's atmosphere within several weeks, although a minor portion might persist for up to a year. This assessment was echoed by Brian Weeden of the Secure World Foundation, who acknowledged the rapid decay of most debris but cautioned about the potential for some fragments to be propelled into higher orbits.

The international response to Mission Shakti was measured, reflecting a balance between acknowledgment of India's technological achievement and concerns over space debris. Patrick Shanahan, the acting United States Secretary of Defense at the time, highlighted the inherent risks associated with ASAT tests, particularly the proliferation of space debris. However, he also indicated that the debris generated by India's test was not expected to pose a long-term threat. The United States Department of State recognized India's commitment to minimizing debris and reiterated its intention to collaborate with India on space security matters. In a gesture of diplomatic goodwill, Russia acknowledged India's declaration that the test was not aimed at any specific nation and extended an invitation to India to join a proposed Russian–Chinese treaty aimed at preventing the weaponization of space.

Mission Shakti not only underscored India's entry into the elite cadre of ASAT-capable nations but also highlighted the broader implications of space weaponization. The test demonstrated India's ability to safeguard its space assets and assert strategic deterrence in an era where space has become a critical theater for national security. Moreover, India's approach to conducting the ASAT test with an emphasis on minimizing space debris set a precedent for responsible behavior in the militarization of space, potentially influencing international norms and agreements.

The evolution of India's ASAT program reflects the dynamic interplay between technological innovation, strategic necessity, and international diplomacy. As nations continue to vie for supremacy in space, India's advancements contribute to the complex mosaic of space security, where the preservation of the orbital environment and the prevention of an arms race in space remain paramount. India's journey from technological development to operational capability in ASAT systems illustrates the broader trends shaping the modern Space Race, where emerging powers are increasingly assertive in securing their interests in the final frontier.

Israel Space Agency

The Israel Space Agency (ISA) was established as a key governmental entity within Israel's Ministry of Science and Technology, tasked with coordinating all national space research programs with both scientific and commercial objectives. This agency, which has grown into a vital part of Israel's strategic and technological advancements, was founded in 1983 by renowned theoretical physicist Yuval Ne'eman. The ISA emerged as a replacement for the earlier National Committee for Space Research (NCSR), which had been established in 1960 to develop the basic infrastructure required for space-related activities. Under the leadership of Ne'eman, and with a focus on self-reliance, the ISA marked Israel's entry into the exclusive group of nations with indigenous space launch capabilities.

Space research in Israel dates back to the late 1950s, with the formation of the NCSR by the Israel Academy of Sciences and Humanities in 1960. Initially focused on fostering academic research, the NCSR was not intended to develop a full-fledged space program. However, by the 1970s, it had established a foundation for space exploration through the creation of necessary research and technological infrastructure. One of its early achievements came in 1961 when Israel successfully launched a two-stage rocket, demonstrating its emerging capabilities.

By the late 1970s, geopolitical tensions in the Middle East, particularly with Egypt and Syria, underscored the need for advanced reconnaissance capabilities. Traditional reconnaissance flights had become increasingly difficult, prompting the Israeli leadership to explore satellite-based solutions. In 1979, a proposal for a national satellite program was put forward, and after a thorough feasibility study, the "Ofeq Program" was initiated in 1982. The program, which aimed to create an observation satellite system, was a bold effort to ensure Israel's autonomy in space technology, without relying on foreign assistance. The development of the satellite's launch vehicle, known as the Shavit rocket, began around this time, involving collaboration between the Israel Military Industries (IMI), Rafael Advanced Defense Systems, and Malam, each responsible for different components of the project.

In January 1983, the Israeli government formally authorized the establishment of the Israel Space Agency. This decision, made during a high-level meeting attended by Prime Minister Menachem Begin and Defense Minister Ariel Sharon, aimed to advance the nation's burgeoning space ambitions. Though Ehud Barak, then-head of military intelligence, advocated for redirecting resources away from the space program to focus on other priorities, his view was ultimately overruled. Defense Minister Moshe Arens strongly supported the continuation of Israel's space efforts, leading to the resumption of development. By 1984, the National Space Knowledge Center had been established in partnership with Israel Aircraft Industries (IAI), and a contract was signed between IAI and the Ministry of Defense to create the necessary infrastructure for launching Israel's first observation satellite.

On September 19, 1988, the launch of the Ofeq-1 satellite marked Israel's entry into the ranks of nations capable of launching satellites into orbit. This milestone made Israel the smallest country to develop such an indigenous capability. The success of the Ofeq satellite series, designed for reconnaissance purposes, was soon followed by the development of other satellite systems, including the AMOS series for communication, EROS for Earth observation, and TechSAR, a radar-based observation satellite developed in cooperation with the Technion, Israel's Institute of Technology.

In parallel with these developments, the ISA pursued international collaboration, recognizing the importance of global partnerships in advancing space technology and research. Over the years, Israel has signed cooperation agreements with numerous space agencies, including NASA (United States), CNES (France), CSA (Canada), ISRO (India), ASI (Italy), and DLR (Germany), among others. These partnerships have enabled Israel to participate in joint missions and to access advanced technology and research.

A guiding vision for the ISA was established in 2005, underscoring the importance of space research for national defense, technological progress, and economic development. The ISA's mission has been to ensure Israel remains among the leading nations in space research and exploration, with the overarching goal of enhancing the country's scientific and technological capabilities. The ISA's vision also emphasizes the need to promote international cooperation, foster ties between Israeli society and space research, and build the human and technological infrastructure required for sustained advancement in space.

As part of its long-term strategy, the ISA's budget saw a significant increase in 2010 to $80 million, aimed at boosting research and development. However, this budget did not cover launch vehicle development, which was funded on a project-by-project basis. One of the ISA's notable cooperative programs is Project Venus, a joint mission with France's CNES, with a budget of $50 million. Another major initiative is the SHALOM mission, a joint hyperspectral satellite mission with Italy's ASI, which has a budget of $116 million.

Today, the Israel Space Agency continues to play a central role in the country's scientific and technological ecosystem, positioning Israel as a significant player in the global space arena, both in terms of reconnaissance and commercial satellite capabilities. Through the vision and persistence of its leaders, and with the support of international partners, Israel has solidified its place as a pioneering nation in space exploration.

In August 2008, the Israeli and United States governments initiated the development of an advanced upper-tier component to Israel's Air Defense Command, known as Arrow 3. This system, designed to intercept ballistic missiles with a reported kill ratio of approximately 99%, was the result of a study conducted between 2006 and 2007. The study determined the need for an upper-tier element within Israel's ballistic missile defense architecture to complement its existing Arrow 2 system. Arieh Herzog, then Director of Israel's Missile Defense Organization (IMDO), outlined that the key feature of this new tier would be an exoatmospheric interceptor, which would be developed jointly by Israel Aerospace Industries (IAI) and Boeing.

The Arrow 3 system required advanced capabilities beyond those provided by the "Green Pine" and "Super Green Pine" radars used with the Arrow 2. These new requirements included longer-range detection, tracking, and discrimination capabilities, enabling more precise targeting of ballistic threats. In addition to the enhanced "Green Pine" radars, the system considered the deployment of airborne electro-optical sensors on high-flying unmanned aerial vehicles and the integration of the AN/TPY-2 radar, operated by U.S. forces and already stationed in Israel. This suite of sensors and tracking systems would provide a comprehensive defense network capable of intercepting threats beyond the Earth's atmosphere.

The Arrow 3, also known as "Hetz 3" in Hebrew, represents a major leap forward in Israel's missile defense capabilities. The missile is designed to intercept ballistic missiles during their space-flight phase, including intercontinental ballistic missiles (ICBMs) that may carry nuclear, chemical, biological, or conventional warheads. Jointly funded and produced by Israel and the United States, the system is overseen by the Israeli Ministry of Defense's Homa Administration and the U.S. Missile Defense Agency. A key feature of the Arrow 3 missile is its ability to perform exoatmospheric interceptions, leveraging a highly agile kill vehicle equipped with divert motor technology. This allows the missile to dramatically change direction mid-flight, enabling it to track and engage incoming targets, including satellites. The missile has a reported flight range of up to 2,400 kilometers (1,500 miles), making it a formidable tool in defending against long-range threats.

In 2011, IAI conducted preliminary tests of the Arrow 3, which included checks on the propulsion system and tracking sensors. On January 23, 2012, the Israeli Ministry of Defense released footage of a successful fly-out test conducted at Palmachim Airbase. During this test, a model of the Arrow 3 interceptor was launched, validating the starting and propulsion systems. Further testing followed, with Boeing officially joining the Arrow 3 development program in January 2012, contributing between 40% and 50% of the production content, including motor cases, power devices, navigation units, and several avionics packages.

A significant milestone was reached on February 25, 2013, when another fly-out test was conducted, pushing the missile to hypersonic speeds and reaching an altitude of 100 kilometers (62 miles), entering space. The test demonstrated the missile's ability to track stars and perform complex maneuvers in space. Subsequent tests in 2014 and 2015 focused on validating the missile's exoatmospheric interception capabilities. By December 2015, Arrow 3 successfully intercepted a target during a complex test that involved discriminating between real and decoy targets, paving the way for the system's low-rate production.

In July 2019, a series of tests conducted at the Pacific Spaceport Complex in Kodiak, Alaska, marked a major achievement for the Arrow 3 system. The system successfully intercepted three simulated enemy rockets, one of which was intercepted outside Earth's atmosphere, demonstrating its ability to engage exoatmospheric targets with precision.

The Arrow 3 missile employs a two-stage interceptor, similar to the Arrow 2, but is based purely on hit-to-kill technology. This method relies on a solid rocket motor equipped with a thrust-vectoring nozzle, which allows the kill vehicle to adjust its course with exceptional precision. The Arrow 3 kill vehicle also uses proportional navigation to line up with its target's flight path, ensuring a body-to-body interception. The system's high agility and advanced divert capabilities reduce the burden on detection and tracking systems, making it a cost-effective and reliable solution for missile defense.

The Arrow 3 system's capabilities extend beyond missile defense, with some experts suggesting it could be used as an anti-satellite weapon. If proven, this would place Israel among a select group of nations capable of destroying orbiting satellites. IAI displayed a full-sized model of the Arrow 3 missile and its kill vehicle at the 2009 Paris Air Show, further showcasing Israel's growing technological prowess in missile defense.

The Arrow 3 system was officially declared operational on January 18, 2017. Since then, it has been integrated into Israel's multi-tier missile defense network, complementing systems like Iron Dome and David's Sling. In 2023, during the conflict between Israel and Houthi forces in Yemen, the Arrow 3 system was used operationally for the first time. On October 31, 2023, an Arrow 2 missile intercepted a long-range ballistic missile launched by Houthi forces, marking the first recorded instance of space warfare, as the interception took place outside Earth's atmosphere. On November 9, 2023, Arrow 3 successfully intercepted another missile aimed at the southern Israeli city of Eilat, demonstrating its critical role in defending against high-altitude, long-range threats.

Beyond its use in Israel's defense, Arrow 3 is poised for export, with Germany among the first international buyers. As part of the European Sky Shield Initiative, Germany is purchasing the system to defend against potential missile threats from Russia. The agreement, valued at $3.5 billion, represents the largest defense deal in Israel's history and is expected to bring Arrow 3 to operational status in Germany by 2025. Other countries, such as Azerbaijan, have also expressed interest in the system, particularly during periods of heightened tension with neighboring Iran.

Through its advanced design and successful operational use, the Arrow 3 missile defense system has proven itself as one of the most sophisticated and capable missile interception technologies in the world. Developed through close cooperation between Israel and the United States, Arrow 3 continues to serve as a cornerstone of Israel's defense strategy, ensuring the protection of its airspace against evolving threats.

Chapter 12 - Militarizing Space

The US Space Force's stance on militarizing space is shaped by the need to protect U.S. interests in an increasingly contested and militarized domain. While the Space Force does not advocate for space becoming an active theater of warfare, it acknowledges the growing threats posed by adversaries and the need to ensure that U.S. assets can operate freely and securely in space. At the same time, the Space Force promotes responsible behavior in space and supports international efforts to establish norms and rules that reduce the risk of conflict, balancing the imperative of defense to maintain space as a functional and peaceful domain.

The US Space Force's stance on the militarization of space is rooted in the recognition that space is a critical domain for national security, as well as an increasingly contested environment where potential adversaries are developing capabilities to challenge U.S. assets. While the Space Force is not advocating for an arms race in space, it acknowledges the need to protect U.S. interests and defend against emerging threats, emphasizing the importance of space as a domain for both military and civilian operations.

From the Space Force's perspective, the militarization of space is not a new phenomenon. Space has long been used for military purposes, primarily through satellites that provide critical functions such as missile warning, communications, navigation, intelligence gathering, and surveillance. What has changed in recent years is the growing concern that space, once seen as a domain of peaceful exploration and use, is becoming increasingly weaponized. Adversaries like China and Russia are developing anti-satellite (ASAT) weapons, electronic warfare capabilities, and cyber tools designed to interfere with or destroy space-based assets. In response to these growing threats, the US Space Force was established to protect and defend the U.S. space infrastructure.

The Space Force's stance on militarization is built on three key principles: deterrence, defense, and resilience. The US Space Force seeks to deter adversaries from attacking U.S. satellites or other space assets by demonstrating that such actions would be met with a strong response. This deterrence strategy is supported by the development of defensive capabilities that ensure U.S. satellites can continue to operate even in the face of interference or attack. Additionally, the Space Force focuses on building resilience into its space systems, ensuring that critical functions such as missile warning and communication can continue even if individual satellites are compromised.

While the Space Force is tasked with protecting U.S. interests in space, it also supports the peaceful use of space. The U.S. remains committed to international treaties such as the 1967 Outer Space Treaty, which prohibits the placement of weapons of mass destruction in space and restricts the use of the Moon and other celestial bodies to peaceful purposes. However, the treaty does not explicitly prohibit the deployment of conventional weapons or the use of space for military operations, which has left room for the development of military space capabilities by various nations.

The Space Force recognizes that the militarization of space is not solely a U.S. initiative but is driven by the actions of other spacefaring nations. China, for example, has rapidly advanced its space capabilities, conducting ASAT tests and launching dual-use satellites that can be used for both civilian and military purposes. Russia has also demonstrated its intent to challenge U.S. space superiority, with tests of space-based weapons and the development of satellite jamming systems. In this context, the Space Force sees its role as vital in maintaining the U.S. advantage in space and preventing adversaries from gaining the upper hand.

However, the US Space Force also understands the risks of escalating tensions in space. There is a growing concern that space could become a battleground in future conflicts, with the potential for space warfare to have catastrophic consequences for both military and civilian operations. The proliferation of space debris from ASAT tests and other military activities threatens the long-term sustainability of space, as debris can damage or destroy satellites and spacecraft indiscriminately. As such, the Space Force advocates for responsible behavior in space and works with international partners to promote norms and rules that reduce the likelihood of conflict.

Regarding its operational stance, the Space Force does not seek to deploy offensive weapons in space but focuses on ensuring that the U.S. can defend its assets and maintain freedom of operation in the domain. This includes developing capabilities to detect and respond to potential threats, improving space situational awareness, and enhancing the ability of satellites to maneuver or evade attacks. The Space Force also collaborates with the intelligence community and commercial space operators to strengthen the overall resilience of U.S. space operations.

In conclusion, the US Space Force's stance on militarizing space is shaped by the need to protect U.S. interests in an increasingly contested and militarized domain. While the Space Force does not advocate for space becoming an active theater of warfare, it acknowledges the growing threats posed by adversaries and the need to ensure that U.S. assets can operate freely and securely in space. At the same time, the Space Force promotes responsible behavior in space and supports international efforts to establish norms and rules that reduce the risk of conflict, balancing the imperative of defense to maintain space as a functional and peaceful domain.

The militarization of space began as an extension of the Cold War rivalry between the United States and the Soviet Union. In the mid-20th century, space exploration was driven not only by scientific curiosity but also by military ambitions. Both superpowers sought to demonstrate their prowess in ballistic missile technology, recognizing that control of outer space could offer a strategic advantage on Earth. As a result, space became a critical frontier for the deployment of military technologies, including communications satellites, reconnaissance systems, and early experiments with ballistic missile defense.

In 1957, the launch of Sputnik 1 by the Soviet Union marked a pivotal moment, signaling the beginning of the Space Race and the global realization of the potential military applications of space. As both nations expanded their capabilities, they increasingly relied on satellites for military purposes, including reconnaissance missions that allowed them to monitor each other's military installations from space. The accuracy and resolution of these orbital systems rapidly improved, sparking concern on both sides about the vulnerability of their military assets to space-based surveillance.

By the late 1960s, both superpowers had established a significant presence in space, with satellites serving as essential tools for military intelligence and arms control verification. Spy satellites became a cornerstone of "national technical means of verification," a term used in arms control treaties to describe the use of space-based systems to monitor compliance with disarmament agreements. As the technology advanced, both nations began developing anti-satellite (ASAT) weapons, researching methods to disable or destroy each other's satellites, including directed-energy weapons and orbital nuclear devices.

One of the most consequential developments in space militarization was the advent of intercontinental ballistic missiles (ICBMs). These missiles, capable of carrying nuclear warheads over vast distances, were designed to travel through sub-orbital spaceflight, using the upper reaches of the atmosphere to extend their range. The destructive potential of ICBMs heightened the strategic importance of space, as military planners sought ways to defend against them. Early American efforts to counter ICBMs included programs like Nike-Zeus, which proposed detonating nuclear missiles in the upper atmosphere to intercept incoming warheads.

In the late 1950s, the United States explored various ambitious military projects aimed at establishing dominance in space. Project A119, for instance, considered detonating an atomic bomb on the Moon as a demonstration of U.S. power. Meanwhile, Project Horizon and the Lunex Project envisioned the establishment of military bases on the Moon by the late 1960s, though these plans were never realized. As the Cold War progressed, the U.S. military continued to pursue missile defense systems, culminating in the Safeguard Program of the 1970s, which was designed to intercept Soviet missiles using anti-ballistic missiles (ABMs). However, the Safeguard system faced significant limitations, including the requirement for nuclear warheads to destroy incoming ICBMs, and it was quickly rendered obsolete by technological advances and arms control agreements.

In 1983, U.S. President Ronald Reagan introduced one of the most controversial initiatives in the history of space militarization: the Strategic Defense Initiative (SDI), also known as "Star Wars." This ambitious program aimed to develop a space-based defense system capable of intercepting and destroying nuclear missiles before they reached the United States. Critics argued that the plan was technologically unfeasible and would provoke an arms race with the Soviet Union. Proponents, however, believed that the economic strain of competing with SDI would hasten the collapse of the Soviet Union, a prediction that ultimately proved true as the USSR disintegrated in 1991.

Throughout this period, the U.S. military's reliance on space grew steadily. In 1985, the United States established U.S. Space Command (USSPACECOM), a unified command dedicated to coordinating military operations in space. Space-based systems played a critical role in modern warfare, providing invaluable support during the 1991 Persian Gulf War. Communications, intelligence, navigation, and missile warning satellites became essential tools for the U.S. military, especially in conflicts across the Balkans, Southwest Asia, and Afghanistan.

By the early 2000s, space had become a central component of U.S. military strategy. In 2002, U.S. Space Command merged with U.S. Strategic Command (USSTRATCOM) as part of a broader effort to streamline military operations. This merger coincided with the increasing militarization of space, as satellite systems became indispensable for providing tactical information to warfighters on the ground.

The final culmination of decades of space militarization came in 2019 with the establishment of the United States Space Force, the world's first independent space force. Initially comprising 8,600 military personnel and 77 spacecraft, the Space Force was tasked with protecting U.S. interests in space, reflecting the growing recognition that outer space had become a new theater of military operations. As technology continues to advance, space remains a crucial domain for national security, with the potential to shape the future of global conflict and cooperation.

Operation Hardtack I, conducted in 1958, was a series of nuclear tests by the United States, part of a broader effort to understand the effects and capabilities of nuclear weapons, especially in high-altitude environments. This operation included three significant high-altitude nuclear tests: YUCCA, ORANGE, and TEAK. YUCCA, detonated on April 28 at an altitude of 86,000 feet, marked a pioneering moment in nuclear testing as the first detonation carried out via a balloon. Its relatively small yield of 1.7 kilotons made it notable primarily for the method of delivery. The subsequent tests, ORANGE and TEAK, conducted on July 31 and August 11, respectively, reached much higher altitudes of 252,000 feet and 141,000 feet. These bombs were delivered by rockets and had yields in the megaton range, showcasing the growing capabilities of the U.S. in nuclear and missile technology.

Following Operation Hardtack I, the United States continued to test nuclear weapons in high altitudes, culminating in the Starfish Prime test of 1962. Conducted as part of Operation Fishbowl, Starfish Prime was detonated over Johnston Atoll at an altitude of 400 km, in the ionosphere. The 1.4 megaton bomb produced one of the most dramatic displays of the Electromagnetic Pulse (EMP) effect ever recorded, with its impact felt as far away as Hawaii, 1,400 km from the blast site. This test, conducted during the height of Cold War tensions, highlighted the devastating potential of nuclear weapons not only for their physical destruction but also for their ability to disrupt electronic systems over vast distances.

The Soviet Union, likewise, sought space dominance through the development of innovative military technologies. Among their notable efforts were the R-36ORB Fractional Orbital Bombardment System (FOBS) and the Polyus orbital weapons platform. The R-36ORB, developed in the 1960s, was a unique intercontinental ballistic missile (ICBM) designed to enter low Earth orbit before deorbiting to strike its target. This system was intended to bypass NORAD's early warning systems by approaching North America over the South Pole, rather than from the more heavily monitored northern approaches. However, the deployment of FOBS was eventually phased out in 1983 following the signing of the SALT II treaty in 1979, which specifically prohibited the deployment of weapons in Earth orbit.

In 1987, the Soviet Union launched the Energia rocket, which carried the Polyus orbital weapons platform. Polyus, although never fully operational, represented an ambitious attempt to establish a space-based defense system equipped with nuclear space mines, recoilless cannons, and a sensor-blinding laser to defend against anti-satellite weapons. Despite its failure to achieve orbit, the Polyus project demonstrated the Soviet Union's commitment to maintaining space superiority through increasingly advanced and imaginative technologies.

The Soviet Union also conducted their own high-altitude nuclear tests to study the effects of High-Altitude Electromagnetic Pulse (HEMP) weapons. One of the most significant of these was Nuclear Test 184, conducted in 1962 at an altitude of 290 km. The resulting EMP caused severe damage to a 1,000 km-long power line in Kazakhstan, an incident that demonstrated the profound vulnerability of even well-protected infrastructure to the effects of high-altitude nuclear detonations. The damage from Test 184 was comparable to the strongest naturally occurring geomagnetic disturbances, illustrating the potential for HEMP weapons to disrupt entire power grids.

As the Cold War drew to a close with the dissolution of the Soviet Union in 1991, the intense competition between the two superpowers in space gradually subsided. However, the militarization of space did not end with the Cold War. In the post-Cold War era, the United States emerged as the world's dominant space power, with technological advancements in reconnaissance, communications, and missile detection systems continuing to evolve. The establishment of the Russian Space Forces in 1992 ensured that Russia maintained its own military presence in space, although its capabilities were diminished compared to its Soviet predecessor.

Throughout the post-Cold War period, the militarization of space centered on three primary applications: reconnaissance satellites, early-warning systems, and missile detection. Spy satellites, first deployed during the Cold War, continued to provide essential intelligence, from high-resolution imaging (IMINT) to communications eavesdropping (SIGINT). These capabilities were critical during peacetime and war alike. Early-warning satellites, which were used to detect nuclear detonations and missile launches, became indispensable tools for ensuring national security. During the Gulf War in 1991, for instance, U.S. early-warning satellites provided Israel with advance notice of incoming Iraqi SCUD missile launches, underscoring the critical role of space-based systems in modern warfare.

The post-Cold War era also witnessed the emergence of new spacefaring nations, challenging the United States' monopoly on space militarization. Countries like China, Japan, and India developed their own space programs, while the European Union worked collectively to create satellite systems that could rival those of the U.S. These developments have made the militarization of space an increasingly complex and multifaceted arena, with no single nation holding a guaranteed position of dominance.

Military satellites play a crucial role in the militarization of space, serving as indispensable tools for reconnaissance, communication, navigation, and missile detection. Since the dawn of the space age, nations have invested heavily in the development of these satellites, which have transformed modern warfare by providing real-time intelligence, secure communications, and precision targeting capabilities. Over the decades, a wide variety of reconnaissance satellites have been deployed by several countries, each designed to gather critical intelligence and monitor global activities.

The United States, a pioneer in the deployment of military satellites, has developed several key reconnaissance systems. These include the Lacrosse/Onyx satellites, which are known for their radar imaging capabilities, allowing them to penetrate through cloud cover and darkness. The Misty/Sirconic series was designed to avoid detection by radar and optical sensors, while the Samos satellites were some of the earliest reconnaissance platforms used for capturing photographic intelligence from orbit. The Quasar series has played a vital role in the field of communications intelligence, and the Vela satellites were developed to detect nuclear detonations, especially in the aftermath of the Partial Test Ban Treaty of 1963, which sought to curb above-ground nuclear testing.

The Soviet Union, in its efforts to match U.S. advancements, also launched numerous reconnaissance satellites. The Cosmos series was a vast program that included hundreds of satellites for a variety of military purposes. The Almaz series, which was crewed, was an ambitious project aimed at placing human-operated reconnaissance platforms in orbit. The Yantar and Zenit satellites were pivotal in the Soviet Union's efforts to obtain high-resolution imagery of strategic targets around the world.

Other nations have also developed their own military satellite programs. The United Kingdom launched the Skynet series for military communications, while France developed the Helios satellites for reconnaissance, although Helios 1B was destroyed. Germany's SAR-Lupe satellites and Italy's COSMO-SkyMed systems provide radar surveillance capabilities, which are vital for military intelligence. Japan, with its Information Gathering Satellites, focuses on both photo reconnaissance and radar imagery, while China has developed the Fanhui Shi Weixing series to enhance its intelligence-gathering capabilities. India has launched the RISAT series, designed for all-weather surveillance using synthetic aperture radar. Israel's Ofeq satellites are renowned for providing high-resolution imagery and radar capabilities. Spain's Paz and Turkey's Göktürk satellites contribute to a growing list of nations leveraging space for military reconnaissance.

In addition to reconnaissance satellites, the Global Positioning System (GPS) is another critical application of space militarization. Initially developed by the United States Department of Defense, GPS has become an essential tool for both military and civilian navigation. The system operates with a constellation of at least 24 satellites in intermediate circular orbit (ICO), ensuring global coverage for determining precise location and providing an accurate time reference. GPS was designed with military objectives in mind, including the enhanced command and control of forces, improved situational awareness, and the ability to accurately target munitions such as smart bombs and cruise missiles.

The first GPS satellite in the current constellation, known as Block II, was launched on February 14, 1989, marking the beginning of the modern GPS system. By 2004, the 52nd GPS satellite had been launched aboard a Delta II rocket. The system is not only used for navigation but also carries nuclear detonation detectors, which form a crucial part of the United States Nuclear Detonation Detection System.

Despite being freely available for civilian use, GPS remains under the control of the U.S. military, with the ability to degrade or deny access to civilian users in times of conflict through jamming or spoofing. This level of control has raised concerns among other nations, particularly in Europe, leading to the development of alternative satellite navigation systems. The European Union has planned the Galileo positioning system to ensure independent access to satellite navigation for both civilian and military purposes. Russia, meanwhile, operates its own independent system known as GLONASS, which functions with a network of 24 satellites deployed in three orbital planes, compared to the four used by the GPS constellation.

China has also developed a regional navigation system called BeiDou, which provides the country with independent navigation capabilities in the Asia-Pacific region. As these systems continue to evolve, they underscore the growing importance of satellite navigation in military operations, providing a strategic advantage in modern warfare.

The militarization of space has evolved rapidly, particularly with the advent of network-centric warfare, a doctrine that integrates advanced communication systems to enhance military effectiveness. This approach relies heavily on high-speed communication networks, allowing real-time data sharing across all branches of the military. The seamless integration of satellite technology has enabled soldiers in battle zones to access satellite imagery, identify enemy positions, and relay coordinates to nearby bomber aircraft or weapons platforms. Commanders, even those stationed miles away, can watch these operations unfold on their screens. The U.S. Department of Defense has been instrumental in developing this capability, working to establish a Global Information Grid, a military-specific network designed to connect all military units and create an efficient information-sharing system. This framework is supported by an intricate web of communication satellites, which provide the necessary infrastructure for real-time battlefield awareness.

Amid these developments, the potential for spaceplanes to play a pivotal role in military operations has garnered attention. The U.S. Space Shuttle, despite its civilian applications, raised concerns within the Soviet Union about its possible military uses. Soviet officials feared that the Shuttle might be capable of delivering nuclear weapons into their territory with minimal warning. Though the Buran program, the Soviet response to the U.S. Shuttle, was not primarily motivated by these fears, the study examining the Shuttle's military potential confirmed the Soviet concerns after Buran's approval.

The development of the NASA X-37 spaceplane, which was later transferred to the U.S. Department of Defense, further highlighted the convergence of military and space technologies. While the X-37's exact military mission remains classified, speculation includes testing experimental reconnaissance and spy sensors designed to endure the harsh environment of space. The X-37 has also been used to test Hall thrusters, which are advanced propulsion systems powered by electricity and xenon. These technologies represent a new frontier for unmanned space vehicles, capable of performing military tasks once reserved for manned aircraft.

Weapons in space became a focal point during the Cold War, with both superpowers developing technologies that could potentially target satellites or other space-based systems. Anti-satellite weapons (ASATs) were designed to attack objects in orbit, and their development spurred international debates about the weaponization of space. One of the earliest instances of space weaponry was the Soviet Union's Salyut 3 space station, which was equipped with a 23mm cannon. This weapon was successfully test-fired at target satellites, demonstrating the possibility of defending space assets. Meanwhile, Russian cosmonauts regularly carried the TP-82 survival pistol on their missions, a multi-functional firearm designed to protect them in case of emergency landings in hostile environments. The TP-82 could fire bullets, shotgun shells, and flares, underscoring the practical, terrestrial concerns that extended into space missions.

The use of high-altitude electromagnetic pulses (HEMP) during the Cold War further illustrated the military's interest in harnessing space-related phenomena. HEMPs, produced by atmospheric nuclear explosions, can disable electronics over vast areas, as demonstrated by the United States' Starfish Prime test in 1962, which caused widespread electrical failures in Hawaii, located over 1,400 kilometers from the blast site. The Soviet Union conducted similar tests, resulting in the shutdown of power lines and telecommunications networks. The catastrophic potential of HEMPs emphasized the need to protect electronic infrastructure from such space-based threats, particularly as militaries became increasingly reliant on advanced technology.

Space warfare, while largely hypothetical, has seen some real-world applications. In the 1980s, the United States conducted a successful test in which an F-15 fighter jet shot down a satellite in orbit. More recently, other nations have demonstrated their ability to destroy satellites. China conducted a successful ASAT test in 2007, destroying one of its defunct satellites, and in 2008, the United States followed suit by destroying a malfunctioning satellite. India joined this select group in 2019 by successfully targeting one of its own satellites, and in 2021, Russia destroyed an old Soviet satellite using a ground-based missile. Although no human casualties have resulted from space warfare to date, the destruction of satellites has raised concerns about the creation of space debris and the potential for escalating conflict in the final frontier.

In response to these developments, international treaties have been established to govern military activities in space. The Outer Space Treaty, signed in 1967, remains the cornerstone of international space law. It prohibits the placement of nuclear weapons or other weapons of mass destruction in orbit or on celestial bodies, and it asserts that space exploration should benefit all humanity. The treaty also holds states responsible for their national space activities, whether conducted by governmental or non-governmental entities. Though comprehensive, the treaty does not ban all forms of weaponry in space, only those involving nuclear or mass destruction capabilities.

In addition to the Outer Space Treaty, other agreements have sought to prevent an arms race in outer space. The Moon Treaty, although signed by several nations, has not been ratified by any major spacefaring powers. It aims to prohibit the use of celestial bodies for military purposes, though it allows military personnel to participate in scientific research. The Limited Test Ban Treaty of 1963 also played a role in curbing the militarization of space by banning atmospheric and high-altitude nuclear tests, thereby limiting the potential for HEMPs.

Efforts to prevent an arms race in space have continued into the 21st century. The Prevention of an Arms Race in Outer Space (PAROS) has been a central issue at the United Nations Conference on Disarmament since 1985. In 2008, China and Russia proposed the Treaty on the Prevention of the Placement of Weapons in Outer Space (PPWT), though the United States opposed the treaty due to concerns about its own space assets. Despite these challenges, the United Nations continues to pass resolutions aimed at preventing the militarization of space, emphasizing the need for cooperation and peaceful exploration of the cosmos.

These treaties reflect the global recognition that space, as a shared domain, requires careful governance to prevent conflict and ensure that its benefits are enjoyed by all humanity. The militarization of space, while inevitable in some respects, remains a topic of intense debate, as nations balance their security needs with the desire to maintain space as a realm for peaceful exploration and scientific discovery.

Following the end of the Cold War and the collapse of the Soviet Union, U.S. defense priorities shifted away from the intense military build-up that had characterized the previous decades. Defense spending was reduced, and space research focused increasingly on peaceful exploration and scientific endeavors. However, amidst this transition, the United States recognized the continuing need for a robust defense posture, particularly in the realm of nuclear deterrence. One significant development in this period was the American military's focus on National Missile Defense (NMD), a system designed to protect the United States from the threat of nuclear blackmail or terrorism by rogue states.

The NMD program is not about placing weapons in space, but its operational concept necessitates intercepting incoming ballistic missiles at high altitudes, often in space. This missile defense system relies on advanced interceptor missiles, which may be launched from land-based or sea-based platforms, to destroy hostile warheads before they reach their targets. The interceptors are engineered to engage the warheads in space, far above the Earth, where the threat can be neutralized without endangering civilian populations.

On December 16, 2002, U.S. President George W. Bush signed a National Security Presidential Directive outlining a plan to deploy operational ballistic missile defense systems by 2004. The following day, the U.S. government formally requested the use of facilities in RAF Fylingdales in England and Thule Air Base in Greenland, both key locations for the NMD program. These facilities were strategically important as they hosted advanced radar systems capable of tracking incoming ballistic missiles, providing crucial early-warning data to support interception efforts.

Despite the administration's enthusiasm for missile defense, the program faced challenges from various quarters. Many scientists voiced ethical concerns, arguing that missile defense could destabilize global deterrence strategies and potentially trigger a new arms race. In addition, the development of the system encountered a series of technical hurdles. Some of these failures during testing phases were widely publicized, raising doubts about the feasibility of the system. However, these setbacks were not unexpected from a technical standpoint, as the complexity of intercepting a missile traveling at hypersonic speeds in space requires advanced technology that naturally undergoes trial and error.

Despite these challenges, the NMD program continued to move forward, albeit at significant cost. The projected budget for the system between 2004 and 2009 was a staggering $53 billion, making it the largest single line item in the Pentagon's budget during that period. The program's scale and expense reflected the U.S. government's commitment to protecting the nation from the evolving threats of nuclear proliferation and missile attacks. In the context of space militarization, NMD represents a defensive strategy that indirectly involves space, using its vast expanse as the stage for intercepting threats before they can cause harm.

The Rods from God

In the 1980s, it was rumored in classified military circles that the US had deployed a space-based weapons system known as Rods from God. Whether true or not, this has never been acknowledged or confirmed.

In 2003, the United States Air Force outlined a conceptual space-based weapons system known as Hypervelocity Rod Bundles. These weapons, often referred to as "rods from God," are designed to deliver devastating kinetic strikes from space. The system envisions satellite-controlled tungsten rods, each 20 feet long and 1 foot in diameter, with the ability to strike any point on Earth. These rods would be released from satellites in orbit and reach impact speeds of Mach 10 upon reentry into the Earth's atmosphere.

The fundamental principle behind this system is the use of kinetic energy to achieve destruction. As the rods fall from orbit, traveling at speeds of approximately 8 kilometers per second (Mach 24) in space and decelerating to around 3 kilometers per second (Mach 8.8) upon atmospheric reentry, they retain an immense amount of kinetic energy. The resulting impact could cause significant damage, comparable to that of a small tactical nuclear weapon, without the accompanying radioactive fallout. These rods are designed primarily as bunker busters, capable of penetrating deeply buried or fortified targets, such as nuclear bunkers, with remarkable precision and force.

Due to the nature of their satellite deployment, the Hypervelocity Rod Bundles system offers a significant advantage in terms of strike timing and flexibility. A constellation of just six to eight satellites in orbit could target any location on Earth within 12 to 15 minutes, a response time much shorter than that of intercontinental ballistic missiles (ICBMs), which also provide early launch warnings to the targeted area. In contrast, the launch of the tungsten rods would be difficult to detect. Their small radar cross-section and the fact that any infrared signature occurs in orbit—away from fixed positions on Earth—make them elusive targets for defensive systems.

The kinetic energy unleashed upon impact is immense. A tungsten cylinder measuring 6.1 meters in length and 0.3 meters in diameter, traveling at Mach 10, would release kinetic energy equivalent to approximately 11.5 tons of TNT upon impact. With a mass exceeding 9 short tons, the destructive potential of these projectiles makes them highly effective for specific applications, such as targeting deeply buried command and control facilities. However, the practical application of such a system would be limited to situations where the unique characteristics of the tungsten rods, such as their capacity for deep penetration and precision targeting, outweigh the logistical and cost challenges. Conventional bombs of similar weight could deliver comparable destructive capability at a lower cost, but they lack the unique strategic advantages of the space-based system.

The design of the rods is critical to their effectiveness. Their elongated shape and substantial mass maximize sectional density, minimizing energy loss due to air friction and enhancing the rods' ability to penetrate hardened or buried targets. Atmospheric reentry, however, presents its own challenges, particularly with regard to the intense heat generated as the rods descend through the atmosphere. While tungsten is highly resistant to heat, non-tungsten components may require special protection to prevent melting or degradation during reentry.

From a defensive standpoint, the system would be extremely difficult to counter. The high velocity of the rods, coupled with their small size and near-invisible launch signature, would make it nearly impossible for current missile defense systems to intercept or neutralize the threat. The only significant challenge to the system's effectiveness would be ensuring the structural integrity of the rods and components during reentry, a problem mitigated by the selection of materials like tungsten.

This concept of hypervelocity rod bundles, colloquially dubbed "rods from God," represents a unique evolution in the militarization of space. While the system remains largely theoretical, it showcases the growing interest in utilizing space for strategic military applications, offering unprecedented strike capabilities with precision, speed, and minimal warning. Such a weapon would add a formidable tool to any nation's military arsenal, revolutionizing how deeply buried or high-value targets are neutralized in future conflicts.

Chapter 13 - The US Space Force (USSF)

In the long haul, our safety as a nation may depend upon achieving "space superiority." Several decades from now, the important battles may not be sea battles or air battles, but space battles, and we should be spending a certain fraction of our national resources to ensure that we do not lag in obtaining space supremacy.
— Major General Bernard Schriever, 19 February 1957
Creating a new military service...would be a dramatic step. Perhaps a "Space Corps" would be a step toward a Space Force. Maybe the Air Force will preempt these dramatic changes by truly becoming the "Space and Air Force."
— Senator Bob Smith, 18 November 1998

The transition from the U.S. Air Force Space Command to the independent U.S. Space Force marked a significant evolution in the U.S. military's approach to space. By creating a separate branch dedicated to space operations, the U.S. recognized the growing importance of space as a warfighting domain and the need for a focused strategy to protect national interests in space. While the transition posed challenges in terms of organizational restructuring and resource allocation, it also presented opportunities for innovation and technological advancement in the space domain.

The United States Space Force (USSF), one of the eight uniformed services of the United States, the space service branch of the U.S. Armed Forces, is one of only two independent space forces in the world, alongside the Chinese People's Liberation Army Aerospace Force. The creation of the USSF reflects a growing recognition of the strategic importance of space in modern military operations.

The transition from the U.S. Air Force Space Command (AFSPC) to the U.S. Space Force (USSF) was a monumental reorganization within the U.S. military, born from the increasing recognition that space was no longer just a supporting element of military operations but a critical domain in its own right. For decades, space operations had been conducted under the Air Force, but the demands of a new era required a rethinking of how the United States approached the protection and advancement of its space capabilities.

*The first 86 Space Force lieutenants
commissioned from the United States Air
Force Academy on 18 April 2020*

The Air Force Space Command, founded on September 1, 1982, was initially tasked with conducting satellite control, space surveillance, missile warning, and managing missile defense operations. For nearly four decades, AFSPC played a key role in overseeing most of the U.S. military's space operations, including launching satellites, monitoring potential threats in space, and ensuring communications across vast distances. Yet, despite the critical nature of these functions, space was just one of many priorities for the Air Force. As technological reliance on satellites for communication, navigation, and missile detection grew, so too did the understanding that space deserved its own focused attention.

With emerging threats from adversarial nations like China and Russia—both of whom were developing anti-satellite weapons and other forms of electronic warfare—the vulnerability of space assets became increasingly apparent. These nations were actively seeking to challenge U.S. dominance in space, and it became clear that in order to maintain a strategic advantage, the United States needed a military branch that was entirely dedicated to space. The idea of space as a distinct warfighting domain, much like land, sea, air, and cyber, gained momentum, particularly under the Trump administration, which pushed for the creation of the U.S. Space Force.

On June 18, 2018, President Donald Trump announced his intent to create the Space Force as the sixth branch of the U.S. Armed Forces. What began as an executive order set into motion a series of legislative and organizational shifts that would ultimately culminate in the formal creation of the U.S. Space Force. Over the following year, discussions and debates within the Department of Defense and Congress shaped the future of this new military branch. The consensus was clear: space operations needed to be centralized and given a higher priority in order to respond to the rapidly evolving landscape of space-based threats.

The formal transition was authorized by the National Defense Authorization Act (NDAA) for Fiscal Year 2020, signed into law on December 20, 2019. With this legislation, the U.S. Space Force was officially established as an independent branch of the military. However, like the U.S. Marine Corps, which operates under the Department of the Navy, the Space Force would remain nested within the Department of the Air Force, allowing for close collaboration while maintaining its distinct identity.

On the same day the NDAA was signed into law, AFSPC was redesignated as the U.S. Space Force. This marked the formal beginning of the transition. AFSPC's structure was rebranded and reorganized into the Space Force, and its commander, General John W. Raymond, was named the first Chief of Space Operations (CSO). As the highest-ranking officer in the Space Force, Raymond was tasked with overseeing the transition, shaping the new branch's mission, and leading its development.

The transition process was complex and occurred in several phases. One of the first key steps was the transfer of personnel. Thousands of individuals who had been working in space-related roles within the Air Force were gradually reassigned to the Space Force. Many of these early personnel had come from existing AFSPC units that were involved in satellite control, space intelligence, and missile warning. As the Space Force matured, it began to develop its own unique personnel structure, recruiting new talent directly into its ranks rather than relying on transfers from the Air Force.

In addition to the transfer of personnel, many space-related units and commands that had once been part of the Air Force were redesignated as Space Force units. For instance, the 14th Air Force was renamed Space Operations Command (SpOC), which became one of the primary operational units of the new Space Force. Further, the establishment of new organizational structures was necessary to streamline space operations and ensure the Space Force functioned as an independent branch. This included the creation of dedicated commands responsible for space training, logistics, and acquisitions.

One of the most significant challenges during the transition was developing a new military doctrine and strategy specific to the Space Force. While space had long been part of the Air Force's strategy, the Space Force needed to establish its own approach, focusing on space superiority, the protection of satellites, and the management of space assets critical to national security. Its responsibilities encompassed everything from ensuring the reliability of global positioning systems (GPS) and missile warning networks to safeguarding secure satellite communications and intelligence operations.

The transfer of space assets followed as part of this reorganization. Satellites, ground stations, space launch operations, and other key space infrastructure, which had previously been under the purview of AFSPC, were moved to the Space Force. These assets, essential for missile warning and space domain awareness—monitoring the objects and activities in Earth's orbit—were now firmly within the control of the new military branch. In parallel, the Space Force focused heavily on developing space technologies, ensuring that U.S. leadership in space would remain unchallenged in the face of growing threats. This involved accelerating the acquisition of new space technologies, improving the resilience of satellites, and advancing space situational awareness.

Despite the many opportunities the transition provided, there were challenges. One of the most significant was shifting the organizational culture from a service within the Air Force to a fully independent branch. This required creating new uniforms, insignia, rank structures, and traditions, while simultaneously building an identity separate from the Air Force. Efforts were made to communicate the importance of the Space Force as a critical, distinct entity within the U.S. military.

Recruiting and retaining the right talent was also critical. The Space Force required highly specialized personnel in areas such as space operations, engineering, cyber, and artificial intelligence. Recruitment strategies were adapted to focus on science, technology, engineering, and mathematics (STEM) fields. Equally important was retaining the experienced personnel who had come from AFSPC and had deep knowledge of space operations.

Financial resources were another critical component of the transition. The creation of new training programs, organizational restructuring, and the acquisition of advanced space technologies all required substantial investments. Ensuring that the Space Force had the funding needed to become fully operational was essential for its success.

Finally, the creation of the U.S. Space Force had global implications. Allies and adversaries alike took notice of this bold move by the United States, and while some expressed concerns about the militarization of space, others saw it as a necessary step in maintaining peace and security in the increasingly contested domain of space. The Space Force would need to navigate these complex geopolitical dynamics, working with international partners while maintaining its mission to protect U.S. interests.

Thus, the transition from AFSPC to the U.S. Space Force was more than just a name change; it was a fundamental shift in how the United States approached space as a military domain, ensuring that the nation remains at the forefront of space innovation, defense, and exploration.

The origins of the United States Space Force can be traced back to the military space programs developed by the Air Force, Army, and Navy during the Cold War. These programs emerged as space became a critical frontier in global competition, driven by the rivalry between the United States and the Soviet Union. The first notable participation of space forces in combat operations occurred during the Vietnam War. Since then, they have played a role in every U.S. military engagement, most notably during the Persian Gulf War, which earned the moniker "the first space war" due to the critical role that space-based systems played in intelligence, communications, and missile guidance.

During the 1980s, military space operations experienced a renaissance. President Ronald Reagan introduced the Strategic Defense Initiative (SDI), a missile defense system that sought to leverage space technology to defend against nuclear attacks. This period also saw the establishment of the Air Force Space Command, a critical step in the formal organization of military space activities. The Air Force Space Command would go on to manage space missions for decades, laying the groundwork for creating a separate space service.

The concept of a U.S. Space Force gained serious traction during the Reagan administration as part of discussions surrounding the Strategic Defense Initiative. However, it wasn't until the late 1990s and early 2000s that Congress began to explore establishing a Space Corps or Space Force in response to increasing global developments in space technology and security. In the 2010s, the idea was reignited as concerns over Russian and Chinese advancements in space-based military capabilities grew. This culminated in the formal establishment of the United States Space Force on December 20, 2019, during the Trump administration.

As an independent branch, the USSF is organized under the Department of the Air Force, making it a coequal sister service to the U.S. Air Force. This structure is seen as an interim step toward establishing a fully independent Department of the Space Force, which a civilian secretary would lead. For now, the U.S. Space Force is led by the Chief of Space Operations, who oversees its missions and operations.

The Space Force's mission, as defined in its statutory responsibilities under 10 U.S.C. § 9081, is to "secure our Nation's interests in, from, and to space." This broad mission encompasses several critical objectives, including providing freedom of operation for the United States in space, conducting space operations, and protecting U.S. interests in the space domain. The Department of Defense further specifies that the Space Force is tasked with deterring aggression in, from, and to space, ensuring continued space operations, and maintaining space superiority.

The Space Force's mission can be broken down into three core functions: Space Superiority, Global Mission Operations, and Assured Space Access.

Space superiority is the ability to maintain control over space, ensuring that the United States and its allies can freely operate in space while denying adversaries the same capability. This involves defending U.S. spacecraft from potential threats and countering enemy spacecraft when necessary. Space superiority also extends to protecting against attacks that rely on space-enabled technologies. To achieve this, the USSF conducts missions related to orbital warfare, electromagnetic warfare, and space battle management, all aimed at maintaining dominance in the space domain.

Global mission operations integrate joint functions across all military domains—land, air, sea, space, and cyberspace—on a global scale. Through space, U.S. military forces and their allies are able to communicate, navigate, and gather intelligence. One of the most critical aspects of global mission operations is providing early warning of incoming missile threats, a function that directly protects U.S. forces and interests on Earth. Space-based systems also enable essential military functions such as missile warning, satellite communications, and the provision of positioning, navigation, and timing services, which are vital to military operations worldwide.

Assured space access ensures the United States can deploy and sustain its assets in space. This includes the launch and maintenance of spacecraft and satellites, as well as the ability to maneuver them in orbit to avoid hazards such as space debris. Assured access to space is a 24/7 mission, ensuring the uninterrupted use of space-based systems for both military and civilian purposes. Key missions supporting this objective include space launch operations, range control, cyber defense of space assets, and maintaining space domain awareness—knowing what is happening in space at all times.

In its relatively short history, the United States Space Force has already become a pivotal element of U.S. national defense, ensuring that the United States remains at the forefront of space exploration, defense, and operations in an increasingly contested and vital domain.

The origins of the U.S. Space Force can be traced back to the aftermath of World War II. In 1945, General Henry H. Arnold, commander of the U.S. Army Air Forces, recognized the emerging importance of advanced technology in future warfare. He tasked General Bernard Schriever with integrating the military with the scientific community to identify and develop technologies that could benefit the newly established U.S. Air Force in a potential global conflict. Schriever, who would later become known as the "father of the Air Force space and missile program," embraced the significance of space in national defense, marking the beginning of the U.S. military's focus on space operations.

During the early Cold War years, the U.S. Army, Navy, and Air Force each initiated separate space and rocket programs. In 1954, the U.S. Air Force took a historic step by establishing the Western Development Division under General Schriever's command, creating the world's first military space organization. A year later, the U.S. Army formed the Army Ballistic Missile Agency (ABMA) under General John Bruce Medaris, with the renowned German rocket scientist Dr. Wernher von Braun leading the agency's technical efforts.

On January 31, 1958, the Army's ABMA launched Explorer 1, the first successful American satellite, which marked the United States' formal entry into space exploration. At this time, space activities were largely managed by the military until the establishment of NASA in October 1958, which shifted the responsibility of space exploration to a civilian agency. However, the military continued to support NASA's missions by providing astronauts, launch vehicles, and recovery operations while maintaining its own focus on national security-related space activities.

By the early 1960s, the U.S. Air Force was recognized as the primary military service responsible for space operations, with the Army and Navy playing supportive roles. Military space efforts during this period centered on developing reconnaissance satellites, weather monitoring systems, communications, and navigation technologies. A significant achievement occurred on August 18, 1961, when the Air Force and the National Reconnaissance Office (NRO) launched the first CORONA reconnaissance satellite. This mission successfully recovered 3,000 feet of film, capturing imagery of 1.65 million square miles of Soviet territory, highlighting the critical role of space in national security.

As the Soviet Union advanced its own space capabilities, the U.S. military recognized the need for human spaceflight as part of its defense strategy. General Curtis LeMay, Chief of Staff of the Air Force, drew parallels between the rapid evolution of aviation during World War I and the potential for space to become a new theater of conflict. He argued that the U.S. must be prepared for the militarization of space, much like air combat had quickly evolved from reconnaissance to warfare.

Despite these strategic concerns, political opposition within the Department of Defense hindered the progress of several key military space initiatives. The Air Force's plans for the X-20 spaceplane, the Manned Orbiting Laboratory, and Blue Gemini were ultimately canceled, as was Project SAINT—a satellite interceptor with capabilities to neutralize enemy satellites—after details of the project were leaked to The New York Times in 1962. However, the Air Force succeeded in fielding Program 437, an anti-satellite weapon system that employed nuclear-armed Thor missiles to target enemy spacecraft.

Throughout the 1960s and 1970s, the military's space operations remained fragmented across different commands. The Air Force established the Air Force Space Command in 1982, recognizing the need for centralized control. This was followed by the creation of the joint U.S. Space Command in 1985, which aligned the Air Force, Navy, and Army space efforts under a unified operational commander. These organizational changes, coupled with President Ronald Reagan's Strategic Defense Initiative, revitalized military space operations and cemented space as a critical domain for national defense.

Space capabilities were first deployed in combat during the Vietnam War, where satellites provided vital weather and communications support for U.S. forces. In subsequent conflicts, such as Operation Urgent Fury in Grenada and Operation El Dorado Canyon in Libya, the military began incorporating space-based intelligence for command and control. However, during the Persian Gulf War in 1991, the full potential of military space forces was realized. More than sixty spacecraft provided critical communications, navigation, missile warning, and weather support to the Coalition forces, leading to a swift and decisive victory. The Gulf War became known as "the first Space War" due to the central role space assets played in shaping the outcome of the conflict.

The use of space-based technologies continued to expand in the 1990s. During Operation Allied Force in 1999, Global Positioning System (GPS)-guided munitions were used for the first time, heralding a new era of precision warfare. In the wake of the September 11, 2001, attacks, U.S. space forces played an essential role in the Global War on Terrorism, supporting operations such as Operation Enduring Freedom in Afghanistan and Operation Iraqi Freedom.

The concept of a separate military service dedicated to space operations dates back to the early days of the space race. In the 1960s, military space activities were briefly consolidated under the newly formed Advanced Research Projects Agency (ARPA) in 1958, with the idea of centralizing space efforts under one organization. However, the move sparked concern among the branches of the U.S. military, with the Air Force, Army, and Navy fearing the creation of a "fourth service" devoted solely to space. This apprehension led to the eventual return of space responsibilities to the individual services.

As space technologies advanced, calls for a dedicated space service grew louder. The first formal recommendation for a U.S. Space Force came in 1982, just before the establishment of Air Force Space Command and before President Ronald Reagan's Strategic Defense Initiative was publicly unveiled. That year, the Government Accountability Office suggested that the U.S. Air Force be reorganized as the U.S. Aerospace Force or a separate U.S. Space Force be created. This proposal prioritized space-based laser weapon development and other strategic space technologies. Speculation also arose that President Reagan might announce the formation of a U.S. Space Force, which accelerated the Air Force's plans to create a dedicated space command.

Despite these early discussions, it wasn't until after the Persian Gulf War that space operations began to gain greater recognition. The Defense Department declared that space had become as critical to national defense as land, sea, and air power. However, many in Congress felt that the military still treated space as an auxiliary to air operations. In 1998, Senator Bob Smith drew comparisons between post-World War I Army aviators and modern Air Force space operators, arguing that space was being underutilized. He called for creating a Space Corps within the Department of the Air Force, which could eventually transition into a fully independent Space Force. This proposal reflected the growing sentiment that space power needed to be more effectively developed and resourced within the Defense Department.

In 2000, Senator Smith spearheaded the creation of a congressional commission to assess the organization of national security space. Known as the Rumsfeld Commission, it released its findings in 2001, warning of a potential "Space Pearl Harbor" if the United States did not strengthen its space capabilities. The commission's most significant recommendation was to establish a Space Corps within the Department of the Air Force, which would eventually evolve into a separate Department of the Space Force. The timeline for this transition was projected between 2006 and 2011.

However, the Air Force strongly resisted the commission's recommendations. Air Force Chief of Staff General Michael E. Ryan publicly stated that an independent Space Force would not be necessary for at least 50 years and only once space operations extended beyond Earth's orbit. In the meantime, some of the commission's suggestions were implemented, such as transferring the Space and Missile Systems Center to Air Force Space Command and establishing the National Security Space Institute. Nonetheless, the broader recommendations were sidelined, particularly in the aftermath of the September 11 attacks, as counterterrorism became the military's top priority. Plans for a Space Corps or Space Force were shelved.

While the U.S. shifted its focus toward counterterrorism, both Russia and China recognized the strategic importance of space. Russia reestablished its independent Space Forces in 2001, and China conducted a destructive anti-satellite missile test in 2007, creating a vast field of space debris. In response, the U.S. Congress convened the Allard Commission in 2008, which echoed the concerns of the Rumsfeld Commission. The Allard Commission noted that the U.S. had become increasingly reliant on space but had done little to secure these assets. Once again, it recommended the creation of a Space Corps or a separate Department of the Space Force.

Despite these recommendations, it wasn't until 2017 that the concept of a Space Corps gained significant traction. Representatives Mike Rogers and Jim Cooper introduced a bipartisan proposal to establish a Space Corps within the Department of the Air Force. Although the proposal faced strong opposition from the Air Force and the Department of Defense, it set the stage for further discussions. In 2018, President Donald Trump publicly endorsed the idea of a Space Force and directed the Department of Defense to begin planning for its establishment.

This led to the release of Space Policy Directive-4, which outlined the creation of a U.S. Space Force as a branch within the Department of the Air Force, with long-term plans for forming a separate Department of the Space Force. In December 2019, Congress passed the necessary legislation. President Trump signed the National Defense Authorization Act, officially establishing the U.S. Space Force as the sixth branch of the U.S. Armed Forces. This fulfilled the long-standing recommendations of the Rumsfeld and Allard commissions, creating a Space Corps within the Department of the Air Force. While this fell short of the fully independent Department of the Space Force originally envisioned, it marked a significant milestone in the United States' recognition of space as a vital domain of military operations.

We are at the dawn of a new era for our Nation's Armed Forces. The establishment of the U.S. Space Force is an historic event and a strategic imperative for our Nation. Space has become so important to our way of life, our economy and our national security that we must be prepared as a Nation to protect it from hostile actions.

— Secretary of Defense Mark Esper, 20 December 2019

The establishment of the U.S. Space Force on December 20, 2019, marked a pivotal moment in the history of military space operations. General Jay Raymond, who was already serving as the commander of U.S. Space Command and Air Force Space Command, became the first official member of the newly formed branch, taking on the role of Chief of Space Operations. The immediate redesignation of Air Force Space Command to United States Space Force symbolized the transition, although the 16,000 personnel under the command technically remained part of the Air Force.

On April 3, 2020, Chief Master Sergeant Roger A. Towberman became the second member of the Space Force, assuming the role of its first senior enlisted leader. Just a few weeks later, on April 18, 2020, the service welcomed its first batch of new officers when 86 cadets from the U.S. Air Force Academy's graduating class were commissioned into the Space Force, expanding its ranks to 88 members. By September 2020, active-duty Air Force space operators began transferring into the new service. A significant milestone occurred on December 18, 2020, when Colonel

Michael S. Hopkins, aboard the International Space Station, became the first astronaut to swear into the Space Force.

While the service began to build its culture and identity, it faced several challenges related to public perception. Due to its futuristic mission and coincidental ties to popular science fiction, the Space Force encountered criticism and jokes that linked it to franchises such as Star Wars and Star Trek. The decision to adopt the Army and Air Force's Operational Camouflage Pattern (OCP) uniform, albeit with blue stitching and a full-color U.S. flag, drew comments comparing its appearance to characters in Return of the Jedi. This was further fueled by the unveiling of the Space Force's service dress uniform, which evoked comparisons to the uniforms of Battlestar Galactica or Starfleet. General Raymond defended the choices, explaining that the combat uniforms were practical for personnel deployed to terrestrial combat zones and that it saved money. Nevertheless, the formal dress uniform, described as "futuristic" in appearance, continued to stir conversation.

The Space Force's insignia also attracted attention, with some critics claiming its delta symbol was a copy of Star Trek's Starfleet logo. In reality, the delta symbol had been in use since 1962 by the Air Force's Ballistic Missile Division, predating Star Trek's television debut by several years. Even Star Trek actor William Shatner weighed in, clarifying that Starfleet's emblem was itself an homage to the military space pioneers who came before.

On March 26, 2020, the Space Force oversaw its first space launch with the successful flight of an Atlas V rocket. This event underscored the operational capacity of the new branch and its commitment to securing U.S. interests in space.

In December 2020, the Space Force introduced the term "Guardian" to describe its personnel, drawing from the heritage of Air Force Space Command's motto, "Guardians of the High Frontier." Alongside this, the service adopted the motto Semper Supra (Always Above) and unveiled its official service song. However, the issue of rank structure sparked debate. Some lawmakers, including Congressman Dan Crenshaw, advocated for the Space Force to adopt naval ranks, aligning it with its celestial mission. Despite the controversy, the Space Force retained the Air Force's rank structure, following the tradition of its predecessor.

As the service formalized its structure, former Air Force Space Command units were integrated into the Space Force in 2020 and 2021. Field commands, equivalent to the Air Force's major commands, were established to manage space operations. The Air Force's space wings and groups were reorganized into mission-focused "deltas," akin to an Army Brigade Combat Team, while space base deltas, similar to Air Force air base wings, took charge of space installations. Two notable bases, Patrick Space Force Base and Cape Canaveral Space Force Station, were among the first to be renamed to reflect their new affiliation.

One of the key objectives of the Space Force was the unification of military space operations, which had previously been divided among the U.S. Air Force, Army, and Navy. The formation of the Space Training and Readiness Delta (Provisional) in 2020 laid the groundwork for the Space Training and Readiness Command, consolidating training units from across the services. Space Systems Command was also established to centralize space acquisitions, although the Air Force Research Laboratory's space R&D efforts remained separate. In 2022, the Navy's Naval Satellite Operations Center and the Army's Satellite Operations Brigade were transferred to the Space Force, marking a historic first—unifying satellite communications under a single service. The Space Force also assumed control of the Army's Joint Tactical Ground Station in 2023, bringing all space-based missile warning operations under its umbrella.

The Space Force's capabilities were tested early in its existence. In January 2020, just weeks after its establishment, the branch played a critical role in providing missile warning during Iran's missile attack on U.S. forces stationed at Al Asad Airbase in Iraq. In 2021, it once again found itself at the forefront of space defense when Russia conducted an anti-satellite missile test, destroying one of its own satellites, Kosmos 1408. The debris from this event threatened the safety of the International Space Station, showcasing the increasing importance of safeguarding space assets in a contested domain.

Through these milestones, the Space Force has evolved from a nascent organization into a crucial component of U.S. defense strategy, navigating both operational challenges and public scrutiny. It continues to solidify its identity as the nation's premier military space service, with an eye toward the future of space as a contested and strategically vital domain.

The United States Space Force is structured to effectively lead, organize, and equip its personnel, known as Guardians, to ensure dominance in the space domain. Its organizational framework is composed of several key components that work together to fulfill its mission: the headquarters staff, field commands, deltas, and squadrons.

At the core of its leadership, the headquarters of the Space Force provides strategic guidance, overseeing all activities related to the force's operations and development. The Space Force is led by the Chief of Space Operations (CSO), a four-star general who serves as the highest-ranking military official in the branch. This position not only guides the Space Force internally but also advises the Secretary of the Air Force, the Department of Defense, and the White House on matters of space security. The Space Force operates under the Department of the Air Force, a structure mirroring the relationship between the U.S. Navy and the U.S. Marine Corps under the Department of the Navy.

The Space Force's leadership hierarchy consists of various senior officers and civilian officials, each responsible for critical domains within the organization. The Chief of Space Operations, currently General B. Chance Saltzman, leads the branch, while General Michael Guetlein serves as the Vice Chief of Space Operations (VCSO). Their combined leadership ensures that the Space Force can respond effectively to emerging threats and challenges in space.

Supporting these key roles is the Chief Master Sergeant of the Space Force (CMSSF), John F. Bentivegna, who serves as the senior enlisted advisor to the Chief of Space Operations. He plays a crucial role in maintaining the welfare, readiness, and development of the enlisted force.

Additionally, a series of deputy chiefs oversee specialized functions essential to the Space Force's operations. These include:

Katharine Kelley, the Deputy Chief of Space Operations for Human Capital (SF/S1), responsible for the recruitment, training, and management of Space Force personnel;

Major General Gregory Gagnon, the Deputy Chief for Intelligence (SF/S2), tasked with overseeing intelligence and counterintelligence activities;

Lieutenant General DeAnna M. Burt, leading operations in cyberspace and nuclear security as the Deputy Chief for Operations, Cyber, and Nuclear (SF/S3/6/10);

Lieutenant General Shawn Bratton, in charge of strategic planning as the Deputy Chief for Strategy, Plans, Programs, Requirements, and Analysis (SF/S5/8).

Each of these individuals is charged with steering their respective areas to ensure that the Space Force can continue to secure and maintain U.S. interests in space. In particular, Lieutenant General Bratton plays a key role in developing long-term strategies for space dominance, while Major General Gagnon ensures that intelligence capabilities are seamlessly integrated with operational planning.

Finally, the Space Force maintains a forward-thinking outlook through positions like the Assistant Chief of Space Operations for Future Concepts and Partnerships, currently held by Air Marshal Paul Godfrey. This role focuses on fostering partnerships with allied nations and private industry while developing future concepts that will shape the Space Force's capabilities in the decades to come.

The organizational structure of the U.S. Space Force is designed to meet the specific operational, training, and equipping needs of the Guardians, the personnel responsible for protecting and advancing U.S. space interests.

At the core of this structure are the field commands, component field commands, Space Force elements, direct reporting units, and field operating agencies, each with a distinct mission that contributes to the overall effectiveness of the Space Force in securing space dominance.

The Space Force's field commands (FLDCOM) are tailored to address particular functions within the service, reflecting the diverse roles required to organize, train, and equip Guardians. Each of the three field commands oversees critical aspects of the Space Force's operations and strategic objectives:

Space Operations Command (SpOC) is tasked with generating and sustaining space warfighting capabilities for combatant commanders. Based at Peterson Space Force Base in Colorado, SpOC ensures that space forces are ready to support U.S. military operations across the globe, providing essential space-based capabilities such as satellite operations, missile warning systems, and space domain awareness.

Space Systems Command (SSC), headquartered at Los Angeles Air Force Base in California, focuses on the development, acquisition, and deployment of resilient space capabilities. SSC is responsible for equipping the Space Force with the tools necessary to operate in contested space environments, from satellite systems to advanced communication and tracking technologies.

Space Training and Readiness Command (STARCOM), also based at Peterson Space Force Base, enhances Guardians' readiness through education, training, and testing. STARCOM is responsible for developing the doctrine and skillsets necessary for Guardians to succeed in both competitive and conflict scenarios, ensuring that the Space Force maintains a technological and operational edge.

The component field commands (C-FLDCOM) are integrated within the broader structure of U.S. military operations, ensuring that space forces are fully incorporated into the planning and execution of multi-domain strategies. Each C-FLDCOM coordinates space operations within specific geographical areas of responsibility, aligning with unified combatant commands:

U.S. Space Forces – Space (SPACEFOR–SPACE) serves as the Space Force component to U.S. Space Command, with its headquarters at Vandenberg Space Force Base in California. This command is responsible for planning, executing, and integrating military spacepower into global operations, covering all U.S. military activities that occur above 62 miles (100 kilometers) from Earth's surface, starting at the Kármán line.

U.S. Space Forces – Central (SPACEFOR–CENT) supports U.S. Central Command (CENTCOM) by integrating spacepower across a region that includes Northeast Africa, the Middle East, and Central and South Asia. It is headquartered at MacDill Air Force Base in Florida.

U.S. Space Forces – Europe and Africa (SPACEFOR–EURAF), based at Ramstein Air Base in Germany, operates under both U.S. European Command and U.S. Africa Command, providing space-based capabilities for operations across Europe, large parts of Asia, the Arctic Ocean, the Atlantic Ocean, and Africa.

U.S. Space Forces – Indo-Pacific (SPACEFOR-INDOPAC) supports U.S. Indo-Pacific Command from Joint Base Pearl Harbor-Hickam in Hawaii, integrating spacepower into U.S. military operations in the Asia-Pacific region.

Space Force elements represent specialized units that focus on supporting the broader intelligence and reconnaissance needs of the U.S. government. The Space Force Element National Reconnaissance Office (SFELM NRO), headquartered in Chantilly, Virginia, works closely with the National Reconnaissance Office (NRO) to design, develop, and maintain America's intelligence satellites. These satellites play a critical role in national security by providing reconnaissance and intelligence capabilities that support both military operations and civilian decision-making.

Direct reporting units (DRU) within the Space Force are centers of expertise and innovation, tasked with rapidly developing and deploying new space capabilities:

The Space Development Agency (SDA), based at The Pentagon in Arlington County, Virginia, focuses on creating resilient, multi-orbit architectures that combine government, commercial, and rapid acquisition technologies. These architectures enhance the Space Force's ability to sense, track, and transport data in space.

The Space Rapid Capabilities Office (SpRCO), located at Kirtland Air Force Base in New Mexico, specializes in fast-tracking the development and deployment of critical space technologies. SpRCO's mission is to address immediate and high-priority space needs, ensuring that the U.S. military remains agile in responding to emerging threats.

The National Space Intelligence Center (NSIC), headquartered at Wright-Patterson Air Force Base in Ohio, provides cutting-edge intelligence and technical expertise to national leaders and military commanders. NSIC delivers detailed analysis and intelligence related to the space domain, helping the U.S. outmaneuver its adversaries by providing insights into the capabilities and intentions of space-faring nations. NSIC's work is crucial in enabling the Space Force to maintain its strategic advantage in space.

Through its field commands, component field commands, elements, direct reporting units, and field operating agency, the U.S. Space Force is structured to address the complex challenges of modern space operations. Each organization plays a vital role in ensuring that the United States retains its leadership in space, both in times of peace and conflict.

The United States Space Force, headquartered in Washington, D.C., maintains an extensive network of bases and installations throughout the United States and internationally. As of 2024, the Space Force operates across 18 states and territories, with 46 bases and installations, reflecting its broad operational scope both domestically and abroad.

Most U.S. Space Force operations are housed at significant installations across the continental United States. Among these are:

Buckley Space Force Base in Aurora, Colorado, which falls under the command of Space Base Delta 2. Buckley plays a vital role in missile warning, space surveillance, and satellite command and control.

Peterson Space Force Base, located in Colorado Springs, Colorado, serves as the headquarters for Space Base Delta 1. It also hosts multiple critical units responsible for space operations, defense, and space surveillance.

Schriever Space Force Base, also in Colorado Springs and under the jurisdiction of Space Base Delta 1, is pivotal in the command and control of Department of Defense satellites, missile warning systems, and global positioning systems (GPS).

Los Angeles Air Force Base in El Segundo, California, functions under Space Base Delta 3. This base is a hub for acquisition, development, and deployment of military satellite systems, highlighting its critical role in space launch and operations.

Patrick Space Force Base in Satellite Beach, Florida, operates under Space Launch Delta 45. Alongside Cape Canaveral Space Force Station, this installation serves as a primary launch site for space missions, including satellite deployments and defense-related launches.

Vandenberg Space Force Base, located in Lompoc, California, is managed by Space Launch Delta 30. This base is responsible for satellite launches and space surveillance, contributing significantly to national security and space exploration efforts.

Cheyenne Mountain Space Force Station in Colorado, a renowned facility, is known for housing part of the United States' missile warning and space surveillance operations. It falls under Space Base Delta 1 and plays a crucial role in space defense and strategic monitoring.

Cape Cod Space Force Station in Sagamore, Massachusetts, is home to the 6th Space Warning Squadron. This squadron focuses on missile detection and warning systems that protect the U.S. and its allies.

Cavalier Space Force Station in Cavalier, North Dakota, houses the 10th Space Warning Squadron, another key player in missile detection and space surveillance.

In addition to its continental U.S. presence, the Space Force manages a series of critical installations located in territories beyond the contiguous states, strategically placed for global space operations:

Clear Space Force Station in Clear, Alaska, is operated by the 13th Space Warning Squadron. This station provides early missile warning and space surveillance, extending U.S. defense capabilities to the Arctic region.

Ka'ena Point Space Force Station in Hawaii serves as a command and control facility for satellite tracking operations, with the 21st Space Operations Squadron ensuring the monitoring of space-based assets.

Maui Space Force Station, also in Hawaii, is managed by the 15th Space Surveillance Squadron. This installation is responsible for space surveillance and debris tracking in Earth's orbit, ensuring the safety and operability of vital space infrastructure.

Pituffik Space Base in Qaanaaq, Greenland, hosts the 821st Space Base Group. This remote outpost serves as a key strategic location for missile warning and space surveillance in the high Arctic, contributing to global defense initiatives.

New Boston Space Force Station in Hillsborough County, New Hampshire, is home to the 23rd Space Operations Squadron. This installation manages satellite control and communication, forming an essential part of the Space Force's global network of space operations.

Each of these installations plays a unique and critical role in maintaining the United States' dominance in space operations, missile detection, satellite control, and space surveillance. The expansive nature of the Space Force's infrastructure is a testament to the growing importance of space in national defense and global security, as well as the increasing reliance on space-based systems for communications, navigation, and intelligence gathering.

As the United States Space Force continues to evolve, the need for its independence as a separate military department has gained increasing attention. Currently, the Space Force operates under the Department of the Air Force, functioning more like a "Space Corps" rather than an independent military branch. This structure aligns with earlier visions from Senator Bob Smith, the 2001 Rumsfeld Commission, and the 2008 Allard Commission, all of which anticipated that the Space Force would begin as a division under the Air Force before eventually achieving full independence. The ultimate goal was to establish a separate Department of the Space Force.

In 2019, Space Policy Directive-4 formalized this progression by directing the Space Force to be initially established under the Department of the Air Force. This directive recognized the growing strategic importance of space in national defense and mandated that the Department of Defense regularly review the status of the Space Force to determine when it would be appropriate to recommend full legislative independence. When that threshold is reached, it is expected that the President will seek congressional approval to establish a separate Department of the Space Force, tasked with taking over all space-related missions from the Air Force.

Since its inception, there have been several discussions about renaming the Department of the Air Force to reflect the inclusion of the U.S. Space Force. Some have proposed changing the name to the Department of the Air and Space Forces, a suggestion echoing earlier efforts in the 1980s to rename the Department of the Air Force to the Department of the Aerospace Force. Similarly, in the 2000s, Congress debated renaming the Department of the Navy to the Department of the Navy and Marine Corps. Despite these proposals, both were met with opposition from the Department of Defense and ultimately failed.

Proponents of the Space Force's independence have also advocated for the creation of an Undersecretary of the Air Force for Space, which was initially included in the Trump Administration's legislative proposal. However, the Senate removed this provision, limiting the Space Force's autonomy. Further calls have been made for the Space Force to have its own public affairs and legal divisions, separate from those of the Air Force, to enhance its operational independence.

When the Space Force was formally established, its mission included the consolidation of military space assets from across the Army, Navy, and Air Force. While the Navy and Air Force quickly transferred their space operations to the Space Force, the Army was more reluctant. The Army transferred its satellite communications and missile warning systems, but calls for a complete transfer of its space forces remain.

A primary point of contention is the transfer of the 1st Space Brigade and the 100th Missile Defense Brigade. These units, integral to the Army's space and missile defense operations, have yet to be fully incorporated into the Space Force. The 100th Missile Defense Brigade operates the Ground Based Interceptor (GBI) system from various locations, including Schriever Space Force Base, Vandenberg Space Force Base, and Fort Greely. The Heritage Foundation and other defense analysts have advocated for transferring these brigades and other space-related Army assets, such as the U.S. Army Space and Missile Defense Command, to the Space Force to create a more unified space defense strategy.

Some within the defense community have also proposed transferring missile defense and intercontinental ballistic missile (ICBM) operations to the Space Force. The Center for Strategic and International Studies has echoed this recommendation, suggesting that consolidating these missions would streamline space operations and improve national security capabilities in space.

However, the Army has continued to maintain a separate cadre of Functional Area 40 space operations officers, responsible for space operations within the Army's framework. Despite this, surveys indicate that over 85% of these officers would willingly transfer to the Space Force if given the option. The RAND Corporation has further supported this shift, conducting studies that recommend transferring the 1st Space Brigade to the Space Force. Such moves would not only reduce redundancies across the services but also strengthen the Space Force's capacity to fulfill its expanding mission.

The United States Space Force has a long-standing and cooperative relationship with the National Aeronautics and Space Administration (NASA), particularly in the realms of military and civil spaceflight. As the primary governmental entities responsible for these sectors, NASA and the Space Force share a deep history that dates back to the early days of the space race. The Space Force's predecessors in the Air Force, Navy, and Army were integral to NASA's initial successes, providing the agency with its first space launch vehicles and many of its early astronauts.

As of today, the Space Force remains a critical partner to NASA, hosting key launch operations at Vandenberg Space Force Base and Cape Canaveral Space Force Station. Occasionally, NASA reciprocates by hosting Space Force launches, particularly for heavy lift missions, at the Kennedy Space Center. The Space Force also continues to support NASA's human spaceflight missions by providing range support through Space Launch Delta 45, ensuring the safety of launches and tracking potential threats to the International Space Station (ISS) and other crewed spacecraft.

The partnership between the Space Force and NASA extends beyond launch operations. They collaborate on space domain awareness, focusing on tracking objects in Earth's orbit, and planetary defense, ensuring preparedness for potential asteroid impacts. Furthermore, Space Force members have the opportunity to become NASA astronauts, a notable example being Colonel Michael S. Hopkins, who was commissioned into the Space Force while aboard the ISS during the SpaceX Crew-1 mission in December 2020.

Another key partner of the Space Force is the National Reconnaissance Office (NRO), a critical agency within the Department of Defense responsible for developing, launching, and maintaining intelligence satellites. The Space Force not only manages the NRO's space launches but also provides around 40% of the agency's personnel. Proposals have been made to integrate the NRO fully into the Space Force, transforming it into a dedicated Intelligence, Reconnaissance, and Surveillance Command within the Space Force, consolidating the national security space infrastructure. This integration would enhance the ability of both entities to maintain American dominance in space.

The Space Force's Space Systems Command, in collaboration with the NRO, manages the National Security Space Launch (NSSL) program. This program is essential for launching highly sensitive government payloads, including satellites that support intelligence and reconnaissance missions. The NRO, under the leadership of figures like its director, Christopher Scolese, has been described as pivotal to ensuring space superiority, providing the U.S. government with unparalleled situational awareness and intelligence, including the most advanced imagery and signals data available.

Additionally, the Space Force maintains a strong relationship with the National Oceanic and Atmospheric Administration (NOAA), co-managing the military's weather satellites. NOAA's Office of Space Commerce also plays a significant role in civilian space situational awareness and space traffic management, ensuring safe operations as commercial space activities expand. The decision to shift space traffic management from the military to the Department of Commerce reflects the growing commercial sector in space and mirrors the Federal Aviation Administration's role in managing air traffic.

The symbolism and culture within the United States Space Force are deeply rooted in both the historical development of spaceflight and the traditions of the U.S. military. One of the most prominent symbols associated with the Space Force is the Delta, a figure that carries both scientific and historical significance.

The Delta symbol has its origins in the late 19th and early 20th centuries, when scientists developed the rocket equation, a fundamental breakthrough that made spaceflight possible. In this equation, the symbol Δv represents the change in velocity, a critical concept in understanding how rockets propel themselves into space. Over time, the Delta took on broader symbolic meanings. During the 20th century, it was adopted to represent aircraft, missiles, and arrows—an emblem of speed, precision, and direction.

The Delta first appeared in military iconography in 1940, when the United States Army Air Forces' 36th Fighter Group incorporated the shape into its shield. This use of the Delta endured, as the emblem continued with the U.S. Air Force's 36th Fighter Wing. Following World War II, the Delta became increasingly associated with the space program. It appeared on the U.S. Air Force and NASA's joint X-15 hypersonic research aircraft, one of the earliest and most iconic vehicles to push the boundaries of space exploration.

In 1962, the Air Force Ballistic Missile Division became the first of many military space organizations to adopt the Delta into its insignia. It symbolized the Air Force's expanding role in space exploration, representing the upward thrust into space and the launch vehicles that would deliver satellites into orbit. Over time, the Delta symbol evolved, and by 2020, it became the centerpiece of the U.S. Space Force's seal and logo. Today, the Delta remains a key element in the insignias of various field commands and unit emblems within the Space Force, symbolizing the force's ongoing mission to protect and dominate in the domain of space.

Members of the Space Force are known as Guardians, a title that reflects the service's unique nature and its storied heritage. This name traces its origins to 1983, when Air Force Space Command adopted the motto "Guardians of the High Frontier." In December 2020, the Space Force formally adopted the title Guardian, giving its members a distinct identity much like the Marines or Airmen. Prior to this, personnel in the Space Force were referred to as space professionals.

The term Guardian emphasizes the protective and strategic role that the Space Force plays in safeguarding the United States' interests in space. As technology evolves and space becomes an increasingly contested domain, Guardians are charged with ensuring the safety of space-based assets and maintaining American superiority in the cosmos.

The motto of the U.S. Space Force, "Semper Supra," meaning "Always Above," encapsulates the service's mission and aspirations. This motto aligns with those of other branches of the U.S. military, such as the Marine Corps' "Semper Fidelis" (Always Faithful) and the Coast Guard's "Semper Paratus" (Always Ready). "Semper Supra" signifies the Space Force's constant vigilance and readiness to operate in space, maintaining superiority over adversaries in this increasingly critical frontier.

The Space Force's official service song also takes its name from the motto, further embedding the phrase into the organization's culture. Together, the title of Guardian and the motto "Semper Supra" reflect the ethos of the Space Force as it embarks on its mission to secure the high ground of space for the United States and its allies.

Through its symbols, titles, and motto, the U.S. Space Force maintains a strong connection to both its military heritage and the cutting-edge technology that defines space exploration. These elements, woven together, provide a powerful cultural identity for the newest branch of the U.S. military as it faces the challenges of the future.

The United States Space Force operates with a diverse array of specialties and roles, designed to support its mission of defending the nation's interests in space. Personnel in the Space Force, both officers and enlisted members, are divided into several career fields, each with distinct responsibilities that contribute to the overall strength and capability of the service. These specialties range from space operations and intelligence to cyberspace operations and engineering, reflecting the highly technical and strategic nature of the Space Force's mission.

Space operators form the Space Force's backbone and represent the service's largest career field. Space Operations Officers (13S) oversee and lead the force's space operations, including areas such as orbital warfare, space electromagnetic warfare, space battle management, and space access and sustainment. These officers plan, organize, and direct space operations programs, formulating policies that guide the Space Force's combat and operational capabilities in space.

Enlisted Space Systems Operators (5S) work in conjunction with officers, conducting day-to-day operations that involve similar disciplines, including orbital warfare and electromagnetic warfare. These enlisted specialists ensure the effective management and control of satellites, space sensors, and other critical assets. Both officers and enlisted members in space operations are awarded the Space Operations Badge upon completing training at the 533rd Training Squadron at Vandenberg Space Force Base, followed by further education at the 319th Combat Training Squadron and the National Security Space Institute.

The Space Force also has a small, highly specialized group of astronauts tasked with commanding, piloting, and operating crewed spacecraft. Space Force astronauts are often assigned to NASA missions, performing duties aboard the International Space Station (ISS) or other spacecraft. They operate Department of Defense payloads and serve as consultants on spaceflight to other government agencies. Space Force personnel must complete NASA's Astronaut Candidate (ASCAN) training at Johnson Space Center to become astronauts. Once they have completed a spaceflight mission, Space Force astronauts are awarded the prestigious observer badge with an astronaut rating, symbolizing their accomplishments in space exploration. Colonel Michael S. Hopkins, who transferred to the Space Force from the U.S. Air Force while aboard the ISS in December 2020, became the service's first astronaut.

Intelligence Officers are critical in the Space Force's ability to maintain situational awareness in space. These officers lead the intelligence, surveillance, and reconnaissance operations, ensuring that the Space Force has accurate and timely information about potential threats. Enlisted intelligence analysts support this effort across multiple fields, including All Source Intelligence, Geospatial Intelligence, Signals Intelligence, Fusion Analysis, and Targeting. These analysts gather, process, and analyze data to provide actionable intelligence for decision-making.

Both intelligence officers and enlisted members undergo specialized training at Goodfellow Air Force Base under the 533rd Training Squadron Detachment 1, followed by advanced education through the 319th Combat Training Squadron and National Security Space Institute. Upon completion, they receive the intelligence badge, signifying their expertise in this critical area.

In an age where the boundaries between physical and digital realms are increasingly blurred, Cyberspace Effects Operations Officers (17S) are essential to the Space Force's mission. These officers oversee the operation of cyberspace weapons systems, satellite communications, and the defense of critical infrastructure from cyber threats.

They lead teams of enlisted cyberspace operators ensuring that the Space Force's digital and space-based assets remain secure.

Training for cyberspace operations is conducted through the Air Force's 81st Training Wing at Keesler Air Force Base, where officers and enlisted members complete the Undergraduate Cyber Training program. They also receive advanced training through the 319th Combat Training Squadron and the National Security Space Institute, earning the cyberspace operator badge upon successful completion.

Officers in acquisition and engineering career fields are tasked with managing and developing the technologies that make Space Force operations possible. Acquisition Managers (63A) oversee the Space Force's procurement process, ensuring that the service acquires the tools and systems needed to carry out its mission effectively. Meanwhile, developmental engineers (62E) specialize in fields such as aeronautical engineering (62EXA), astronautical engineering (62EXB), computer systems engineering (62EXC), electrical/electronic engineering (62EXE), and mechanical engineering (62EXH). Human factors engineers (62EXI) focus on optimizing the interaction between human operators and space systems.

Space Force engineers receive rigorous education at institutions such as Defense Acquisition University and the U.S. Air Force Flight Test Engineer course. They play a pivotal role in developing cutting-edge technologies that support both military and civilian space activities.

Space Force personnel are recognized for their expertise and achievements through a variety of badges, each symbolizing a specific area of specialization and accomplishment. For example, the Space Operations Badge is awarded to those who complete the rigorous training required to manage and execute space operations. Similarly, Space Force astronauts receive the observer badge with astronaut rating after successfully completing a mission in space. Cyberspace operators and intelligence analysts are awarded their respective badges upon completing specialized training in their fields.

The United States Space Force (USSF) is built on a foundation of seven core spacepower disciplines, each critical to maintaining the nation's dominance in space. These disciplines form the backbone of Space Force operations, ensuring that both the United States and its allies retain freedom of access to space while denying that advantage to adversaries.

Orbital Warfare involves expertise in orbital maneuvers, alongside offensive and defensive capabilities, to ensure the uninterrupted operations of U.S. and coalition space forces. Personnel in this discipline are trained to safeguard freedom of movement in space, leveraging their skills to provide critical capabilities to the Joint Force while simultaneously preventing adversaries from exploiting space for their own purposes.

Space Electromagnetic Warfare centers on the strategic use of the electromagnetic spectrum. Space Force personnel must be adept at controlling and maneuvering within this spectrum, using non-kinetic methods to disrupt enemy communications and maintain secure U.S. space operations. This discipline requires deep understanding of how adversaries utilize the electromagnetic spectrum and the ability to deny them access while maintaining the functionality of allied systems.

Space Battle Management is the discipline that focuses on command and control of space assets. Personnel in this field are responsible for identifying hostile entities and making rapid decisions to protect space-based missions. Their expertise enables the identification of threats, the direction of defensive measures, and the coordination of space assets to ensure mission success in a constantly evolving threat environment.

Space Access and Sustainment encompasses the logistics, processes, and support mechanisms necessary to sustain operations in space. This discipline ensures that space assets are not only launched but also maintained and resourced throughout their mission lifetimes, securing the long-term presence of U.S. spacepower in orbit.

Military Intelligence plays a crucial role in space operations, as intelligence-led missions allow the Space Force to anticipate and counter threats. Personnel working in this discipline collaborate with the broader Intelligence

Community to gather surveillance and reconnaissance data, providing critical insights into adversary capabilities and movements in the space domain.

Engineering and Acquisition ensures that the U.S. Space Force remains technologically superior. This discipline involves forming partnerships with other national security organizations, commercial enterprises, and academic institutions to develop cutting-edge capabilities that can defend space assets. Engineers and acquisition specialists play a vital role in equipping the warfighter with the most advanced tools available.

Cyber Operations underpins all Space Force activities by defending the global networks upon which space operations rely. Personnel are trained to protect these networks from cyberattacks, ensuring the integrity of U.S. space assets and systems. They also possess the capability to engage in offensive cyber operations, disrupting adversarial networks and preventing attacks on U.S. space infrastructure.

The U.S. Space Force, as the youngest branch of the U.S. Armed Forces, has adopted a rank structure similar to its predecessor, the U.S. Air Force, but with unique elements reflective of its mission and identity in space. Officers in the Space Force are responsible for strategic planning, managing personnel, and overseeing critical space operations. These leaders are commissioned through three primary avenues: the U.S. Air Force Academy, the Air Force Reserve Officer Training Corps (ROTC), and the Air Force Officer Training School (OTS).

The United States Air Force Academy, located in Colorado Springs, is regarded as the premier commissioning source for Space Force officers. Approximately 10% of each graduating class commissions into the Space Force, while the rest join the Air Force. The Academy has a deep-rooted history with space operations, establishing the world's first Department of Astronautics in 1958. It operates the Cadet Space Operations Squadron, responsible for managing the FalconSAT satellite program, and offers specialized space programs such as the Azimuth program and the i5 Squadron. As of 2023, the Academy offers two space-related majors and a space warfighting minor, further solidifying its role in preparing future space leaders.

Space Force officers can also commission through the Air Force ROTC program, which is available at over 1,100 colleges and universities nationwide. Like the Academy, ROTC commissions officers directly into either the Air Force or Space Force, offering a flexible path for students seeking military careers while pursuing higher education.

The final commissioning path is through the Air Force Officer Training School (OTS). The first two Space Force officers graduated from OTS in October 2020, and by March 2023, the school had graduated its first all-Space Force flight. This route allows individuals who have already earned a college degree to receive the necessary military training before becoming Space Force officers.

In terms of advanced education, the Space Force offers a variety of developmental programs to its officers, including partnerships with the Paul H. Nitze School of Advanced International Studies at Johns Hopkins University. Officers can also attend specialized schools such as the National Security Space Institute, Air Force Institute of Technology, U.S. Air Force Weapons School, and the Space Test Course. These institutions provide critical training in space operations, acquisition, and strategic studies, ensuring that Space Force officers remain at the forefront of spacepower innovation.

Enlisted personnel play an equally vital role in Space Force operations. Like their officer counterparts, enlisted Guardians undergo rigorous training and education to support the service's mission. Enlisted personnel begin their journey by completing Basic Military Training at Joint Base San Antonio, a program modeled after Air Force basic training but enhanced with Space Force-specific curricula. In December 2020, the first group of enlisted Guardians graduated from Basic Military Training, marking the beginning of a new chapter for the service.

Once enlisted, members of the Space Force are automatically enrolled in the Community College of the Air Force (CCAF), where they earn associate degrees in applied sciences. This program is designed to provide enlisted

Guardians with the technical knowledge needed to excel in space operations, including fields such as cyber defense, engineering, and intelligence.

Advanced professional military education for enlisted Guardians is conducted at the Forrest L. Vosler Non-Commissioned Officer Academy, located under Space Training and Readiness Command. Additional opportunities include courses offered by the National Security Space Institute and the Space Test Course, providing specialized training in space operations and technology.

The rank insignia of the U.S. Space Force reflects its identity as the sixth branch of the U.S. Armed Forces. The design for enlisted insignia centers around a hexagon, symbolizing the Space Force's status as the sixth military service. The delta, a key symbol of the Space Force, is featured prominently in the insignia, representing the service's focus on space operations.

For noncommissioned officers, the insignia incorporates the "Delta, Globe, and Orbit" design, representing the interconnected nature of space, Earth, and the Space Force's mission. The stripes for specialists 2 through 4 are based on an early proposal for Air Force ranks known as "Vandenberg stripes," a nod to the Space Force's origins within the Air Force.

Senior noncommissioned officers wear insignia topped with "orbital chevrons," representing the different levels of space operation: low Earth orbit for master sergeants, medium Earth orbit for senior master sergeants, and geosynchronous orbit for chief master sergeants. These orbital chevrons signify the increasing levels of responsibility placed on senior enlisted personnel.

The Chief Master Sergeant of the Space Force is the highest enlisted rank and is symbolized by a "Delta, Globe, and Orbit" emblem encased within a hexagonal wreath. This design reflects the Chief Master Sergeant's role as the senior enlisted advisor to the Chief of Space Operations and the Space Force's most senior noncommissioned officer.

The United States Space Force, as the newest branch of the U.S. military, is in the process of developing its distinct uniforms to reflect its unique identity and mission. These uniforms include the mess dress, service dress, and physical training attire. Until these designs are finalized, Space Force personnel, known as guardians, continue to wear modified versions of the U.S. Air Force's uniforms. The modifications include Space Force insignia on coats and shirts, "Delta, Globe, & Orbit" buttons replacing the traditional "Hap Arnold Star & Wings," and Space Force cap badges instead of Air Force badges. Additionally, enlisted guardians wear Space Force-specific ranks and replace the standard Circle U.S. lapel insignia with a hexagonal version. These temporary uniform changes signify the transition of the Space Force toward a distinct identity within the U.S. Armed Forces.

The primary uniform for Space Force personnel remains the Operational Camouflage Pattern (OCP) uniform, adopted from both the U.S. Air Force and U.S. Army. However, the Space Force distinguishes itself with "space blue" thread for ranks and badges. Guardians also wear a full-color American flag patch on their left sleeve and unique, full-color Space Force patches on their uniforms, further emphasizing their connection to both space operations and national service.

A significant milestone in the development of the Space Force's identity was the unveiling of its official service dress uniform at the 2021 Air, Space, and Cyber conference, hosted by the Air & Space Forces Association. This blue and gray uniform draws inspiration from the vastness of space, with its dark blue color reflecting the space environment. The six buttons on the uniform symbolize the Space Force's status as the sixth branch of the U.S. military. The Physical Training Uniform was also introduced in September 2021, with an expected release date of early 2024 for physical training gear and late 2025 for the service dress uniform.

Space Force cadets at the U.S. Air Force Academy similarly wear modified uniforms, maintaining the overall design of Air Force cadets but featuring a platinum sash in place of the traditional gold sash. This distinctive element sets them apart and signifies their role in the Space Force.

In terms of awards and decorations, the Space Force shares many of the same honors as the Air Force, reflecting their shared organizational heritage under the Department of the Air Force. On November 16, 2020, Secretary of the Air Force Frank Kendall III approved renaming several key Air Force awards to include the Space Force. These changes affect the Air Force Commendation Medal, Air Force Achievement Medal, Air Force Outstanding Unit Award, and other notable distinctions, all of which were modified to reflect "Air and Space" rather than solely the Air Force. The Air Force Combat Action Medal was similarly altered, and the Air Force Special Duty Ribbon was renamed the Developmental Special Duty Ribbon to account for the distinct roles within both services.

One of the most notable ongoing developments is the creation of the Space Force Good Conduct Medal, which was approved on August 30, 2023. This medal will replace the Air Force Good Conduct Medal for enlisted Space Force personnel, further emphasizing the Space Force's independent identity. Additionally, discussions in Congress have included the potential renaming of the Airman's Medal, awarded for non-combat heroism, to the Air and Space Force Medal, paralleling similar honors in the Navy and Marine Corps.

The United States Space Force, the newest branch of the U.S. military, has rapidly developed a range of advanced spacecraft and space systems to assert its dominance in space and ensure the security of American interests. These space-based assets support vital missions, from secure communications and environmental monitoring to missile detection and space surveillance. Together, they form the backbone of the Space Force's operations, enabling the U.S. to maintain a strategic advantage in the increasingly contested domain of outer space.

One of the most critical systems operated by the Space Force is the Advanced Extremely High Frequency (AEHF) constellation, managed by Space Delta 8. With six satellites in operation, AEHF provides secure, global, and jam-resistant communication links for U.S. military operations, even in the most hostile environments. These satellites are vital to ensuring continuous command and control across all military branches, enabling decision-makers to stay connected and maintain operational superiority anywhere in the world.

Similarly essential to the Space Force's communication capabilities is the Defense Satellite Communications System (DSCS), also operated by Space Delta 8. Like AEHF, DSCS supports global military operations, providing the U.S. military with high-capacity satellite communications. Alongside these systems, Space Delta 8 also manages the Milstar system and the Wideband Global SATCOM (WGS) constellation, which bolster the U.S. military's ability to communicate reliably across the globe. WGS, with its ten satellites, represents the highest-capacity military communications system, supporting everything from day-to-day operations to large-scale joint military exercises.

In addition to these communication systems, the Space Force oversees a range of spacecraft focused on environmental monitoring and space surveillance. Space Delta 2, for instance, operates the Defense Meteorological Satellite Program (DMSP), which provides critical weather data to support military planning and operations. Monitoring atmospheric and space weather phenomena, the DMSP helps mitigate potential disruptions to satellite systems, communications, and other space-based assets. Space Delta 2 is also responsible for the Electro-Optical/ Infrared Weather System – Geosynchronous (EWS-G), which offers continuous environmental monitoring from a geosynchronous orbit, further enhancing the military's ability to predict and respond to weather patterns that could affect operations on Earth and in space.

Space Force's commitment to space situational awareness—tracking and monitoring objects in space—falls under the jurisdiction of Space Delta 9. This unit operates several key systems, including the Advanced Technology Risk Reduction (ATRR) satellite and the Geosynchronous Space Situational Awareness Program (GSSAP), both designed to track satellites and debris in Earth's orbit. GSSAP, with its six satellites, operates in geosynchronous orbit, providing vital data that helps protect both U.S. and allied space assets from potential collisions or interference. Another important asset managed by Space Delta 9 is the Space Based Space Surveillance (SBSS) satellite, which plays a crucial role in monitoring the ever-growing population of objects orbiting Earth.

Perhaps the most versatile spacecraft under Space Delta 9's command is the X-37B Orbital Test Vehicle, an unmanned spaceplane designed for long-duration missions. This spacecraft conducts a variety of experiments and tests new technologies in low Earth orbit, pushing the boundaries of spaceflight and contributing to the development of future space systems. The X-37B's ability to remain in orbit for extended periods makes it a unique and valuable asset in the Space Force's efforts to expand the U.S.'s capabilities in space.

A cornerstone of the Space Force's national security mission is its missile warning systems, managed by Space Delta 4. Chief among these systems is the Defense Support Program (DSP), which detects missile launches across the globe through advanced infrared sensors. This early warning capability is crucial to national defense, providing the U.S. with the time needed to prepare and respond to potential threats. Complementing DSP is the Space-Based Infrared System (SBIRS), another advanced missile detection network. With seven satellites, SBIRS not only offers missile warning but also contributes to missile defense, battlespace awareness, and technical intelligence, making it an indispensable asset for both defensive and strategic operations.

Completing the picture of the Space Force's space systems is the Global Positioning System (GPS), one of the most widely used and indispensable satellite systems in the world. Operated by the Positioning, Navigation, and Timing Integrated Mission Delta (PNT IMD), GPS provides precise positioning, navigation, and timing services to military and civilian users alike. With a constellation of 32 satellites, GPS has revolutionized how the world navigates, coordinates operations, and conducts business, proving vital not only to military operations but also to modern society at large.

As the U.S. Space Force continues to grow and evolve, these spacecraft and systems will remain at the forefront of its mission to secure space for American interests. Whether through ensuring reliable communication, providing early warning of missile launches, or maintaining situational awareness of the space environment, the Space Force's fleet of satellites and space vehicles is essential to maintaining the U.S.'s strategic advantage in space. These systems, operated by the Space Force's specialized Space Deltas, are the foundation upon which the nation's security in the space domain rests, safeguarding the interests of the U.S. and its allies in an ever-more contested and congested space environment.

The United States Space Force operates a sophisticated network of space systems designed to maintain security, surveillance, and missile defense in space. These systems are integral to the nation's efforts to monitor potential threats and ensure the safety of its assets in space.

One of the key space systems in the U.S. Space Force's arsenal is the AN/FPS-85 radar. This radar, operated by Space Delta 2, is dedicated to space surveillance. Located at Eglin Air Force Base in Florida, it is the largest radar in the world dedicated to tracking objects in space, capable of detecting and cataloging thousands of satellites and debris in Earth's orbit.

Complementing the AN/FPS-85 is the C-Band Space Surveillance Radar System, also operated by Space Delta 2. This radar system is used to detect and track space objects, providing critical data for maintaining an updated catalog of objects in orbit. The C-Band system is vital in ensuring accurate information on the increasing number of objects, both natural and man-made, that travel through space.

Another essential component of the U.S. Space Force's space surveillance capability is Cobra Dane, a radar stationed in Alaska under the control of Space Delta 4. Cobra Dane plays a dual role, contributing to both missile defense and space surveillance. Its advanced capabilities enable it to track ballistic missile launches and satellites, providing data that supports U.S. defense strategies.

The Ground-Based Electro-Optical Deep Space Surveillance (GEODSS) system, also operated by Space Delta 2, utilizes powerful telescopes to track objects in deep space. These telescopes, located in New Mexico, Hawaii, and Diego Garcia, provide optical data on space objects, helping to track satellites and debris that are too far away for radar systems to detect. GEODSS is essential for keeping tabs on objects in geosynchronous orbit, which is critical for communication and weather satellites.

The Long Range Discrimination Radar (LRDR), under Space Delta 4, enhances the U.S. missile defense posture. Situated in Alaska, this state-of-the-art radar system detects incoming missiles, while also contributing to space surveillance efforts. Its ability to discriminate between threatening and non-threatening objects makes it a cornerstone of both missile defense and space monitoring.

At the heart of the nation's early warning systems is the Perimeter Acquisition Radar Attack Characterization System (PARCS), also operated by Space Delta 4. Located in North Dakota, PARCS provides continuous surveillance of objects in space, as well as missile warning capabilities. This radar system has been operational since the Cold War, proving its lasting importance in U.S. defense infrastructure.

The Space Surveillance Network is bolstered by the Space Fence, operated by Space Delta 2. This advanced radar system, based on Kwajalein Atoll in the Marshall Islands, represents a leap forward in space object tracking. The Space Fence can detect much smaller objects than its predecessors, allowing the U.S. Space Force to monitor more debris and track objects with greater precision.

Another notable tool in space surveillance is the Space Surveillance Telescope, also operated by Space Delta 2. This telescope, positioned in Western Australia, enhances the Space Force's ability to detect small, fast-moving objects in space. Its unique location and advanced technology make it a critical asset in the growing need for space domain awareness.

Rounding out the missile warning and space surveillance systems is the Upgraded Early Warning Radar (UEWR), controlled by Space Delta 4. These radars, positioned in key locations such as Greenland, the United Kingdom, and Alaska, play a vital role in detecting missile launches while also tracking space objects. The UEWR is essential for early missile detection and space situational awareness, ensuring that the U.S. can respond quickly to threats from space or missile attacks.

In addition to these advanced radar and surveillance systems, the U.S. Space Force also oversees a range of space launch vehicles used for deploying satellites and other payloads into orbit. These vehicles are contracted from leading aerospace companies.

The Atlas V, developed by United Launch Alliance, is a reliable medium-lift launch vehicle used for missions that require precise deployment of payloads. It has been a mainstay in U.S. space launches for over two decades, supporting both military and civilian missions.

Rocket Lab's Electron rocket is a small-lift launch vehicle designed for rapid, cost-effective launches of small satellites. Its agile design allows it to deliver payloads into low Earth orbit, making it a key player in expanding the capabilities of small satellite constellations.

SpaceX, with its Falcon 9 and Falcon Heavy launch vehicles, provides medium to heavy-lift capabilities. The Falcon 9, known for its reusable first stage, has revolutionized the cost and frequency of space launches. It is capable of carrying payloads to the International Space Station, launching satellites, and conducting interplanetary missions. The Falcon Heavy, on the other hand, is one of the most powerful rockets in operation, able to carry heavy to super-heavy payloads. It is designed for missions requiring the deployment of large payloads or those destined for deep space.

Northrop Grumman's Pegasus is a unique air-launched small-lift launch vehicle. It is deployed from an aircraft, making it flexible and capable of launching from virtually any location. Pegasus is often used for launching small satellites, providing a quick-response solution for deploying payloads into orbit.

Together, these space systems and launch vehicles form the backbone of U.S. space operations, ensuring the nation's security, expanding its presence in space, and maintaining its leadership in space exploration and technology. The capabilities of these systems reflect the evolving nature of space as a strategic domain, one that is increasingly crowded, competitive, and critical to national defense.

The United States Space Force, while a relatively new branch of the military, has embarked on a comprehensive modernization effort, reflected in its rapidly growing budget and ambitious technological programs. In its early years, the Space Force saw significant increases in funding to support its development and operational needs.

In 2020, the Space Force's initial budget was a modest $40 million, marking the beginning of its operational and strategic foundation. By 2021, this amount had ballooned to $15.34 billion, with the majority allocated to operations, maintenance, procurement, and research. By 2023, the total enacted budget had surged to over $26.28 billion, underscoring the urgency of modernizing and expanding U.S. space capabilities. The largest portions of this budget have been consistently dedicated to Research, Development, Test, and Evaluation (RDT&E), which alone reached over $16.63 billion in 2023 and is projected to surpass $19.55 billion in 2024.

The Space Force's modernization is focused on developing and fielding cutting-edge space systems and technologies. Among these, the Deep Space Advanced Radar Capability (DARC) stands out. DARC will consist of three radar sites strategically located in the United States, Europe, and the Indo-Pacific. This system is designed to track objects in geosynchronous orbit (GEO), ensuring that the U.S. has real-time knowledge of the space environment in one of the most critical regions for satellite operations.

The U.S. Space Force is also extending its surveillance efforts beyond geosynchronous orbit into the cislunar domain, the space between Earth and the Moon, with the development of the Oracle spacecraft. This craft, developed by the Air Force Research Laboratory, will be capable of tracking objects in this vast region. Oracle will operate from a point of gravitational stability between Earth and the Moon, known as a Lagrange point, where it will use advanced sensors to monitor activity in this area. This initiative supports NASA's Artemis program, which aims to return astronauts to the Moon. Oracle's capabilities also align with planetary defense efforts, as it will help track potentially hazardous near-Earth objects (NEOs).

In parallel, the Space Force is investing in pioneering energy technologies. The Arachne spacecraft is a key component of the Space Solar Power Incremental Demonstrations and Research (SSPIDR) Project, an Air Force Research Laboratory initiative aimed at developing a space-based solar power system. Arachne will demonstrate advanced technologies for collecting solar energy in space and transmitting it to Earth via radio-frequency beams. The ability to beam power directly to forward operating bases could reduce the military's reliance on vulnerable supply lines that transport fuel, thus providing a strategic advantage. As this technology matures, it could eventually be extended to civilian applications, much like how GPS originated as a military project before becoming a global utility.

Two additional experiments, SPIRRAL (Space Power InfraRed Regulation and Analysis of Lifetime) and SPINDLE (Space Power Incremental Deployable Experiment), are also part of the Space Force's efforts to develop and demonstrate space-based solar power transmission, further exploring the viability of space-based energy systems.

Another advanced initiative by the Air Force Research Laboratory is the Navigation Technology Satellite-3 (NTS-3). This satellite, which will operate in geosynchronous orbit, is part of the broader effort to enhance the Space Force's Global Positioning System (GPS) constellation. NTS-3 will test innovative techniques to mitigate interference and increase the resiliency of positioning, navigation, and timing (PNT) systems. As a Vanguard program, NTS-3 represents a potentially transformative leap in satellite navigation technology, improving PNT services for military, civil, and commercial users.

The Space Force is also exploring rapid global logistics with its Rocket Cargo program, another Vanguard initiative. Through this program, the Space Force aims to lease commercial space launch services capable of delivering military cargo anywhere in the world. SpaceX's Starship is one of the potential contenders for this program, offering the ability to transport up to 100 tons of cargo across vast distances at unprecedented speeds. If successful, the Rocket Cargo program could revolutionize military logistics, allowing the U.S. to deploy equipment and supplies in a matter of hours to conflict zones or disaster areas.

These modernization efforts, coupled with the Space Force's increasing budget, underscore its pivotal role in maintaining U.S. dominance in space. As space becomes an ever more critical domain for military operations, communication, navigation, and global defense, the Space Force is rapidly evolving to meet the challenges of this new frontier. Its investments in radar, solar power, navigation, and logistics are laying the groundwork for a future where space is not just a strategic asset, but a fully integrated theater of operations.

Epilogue

The development and deployment of anti-satellite (ASAT) weapons have long been a focal point in the strategic calculations of major powers, particularly during the height of the Space Race. While the theoretical capability of one nation to disable another's satellites in times of conflict presents a significant threat to military operations, the practical challenges of executing such attacks have proven formidable.

Intercepting orbiting satellites is far from straightforward. Although there have been instances of successful interceptions at low Earth orbits, several factors complicate these endeavors. Military satellites often employ defensive measures, such as altering their orbital inclinations, to evade detection and interception. The rapid lateral movement of satellites, combined with the time required for an interceptor to ascend to the necessary altitude and adjust its trajectory, makes precise targeting exceptionally difficult. For example, US intelligence, surveillance, and reconnaissance (ISR) satellites typically orbit at approximately 800 kilometers (500 miles) above Earth, traveling at speeds of about 7.5 kilometers per second (4.7 miles per second). In a hypothetical conflict between the United States and China, a Chinese intermediate-range ballistic missile would need to accurately compensate for the satellite's movement over the 1,350 kilometers (840 miles) it covers during the three minutes required to reach its target altitude. Even if such an ISR satellite were successfully neutralized, the United States maintains a robust network of both crewed and uncrewed ISR aircraft capable of conducting missions from distances that keep them out of range of Chinese land-based air defenses.

Higher-altitude satellites, such as those forming the Global Positioning System (GPS) constellation and various communication satellites, orbit at altitudes of 20,000 kilometers (12,000 miles) and 36,000 kilometers (22,000 miles), respectively. These altitudes place them beyond the reach of solid-fueled intercontinental ballistic missiles, which limits the feasibility of direct ASAT attacks using such weaponry. While liquid-fueled space launch vehicles possess the capability to reach these higher orbits, their longer launch times and vulnerability to ground-based attacks before multiple launches can be conducted in rapid succession present significant obstacles. The GPS system, consisting of approximately 30 satellites, is designed with redundancy in mind; at any given time, at least four satellites across six orbital planes are accessible to receivers. Disrupting the GPS network would require disabling a minimum of six satellites, a challenging and resource-intensive task. Even in the event of a successful attack, the resulting signal degradation would last no longer than 95 minutes. Moreover, backup inertial navigation systems (INS) and laser-guided weaponry would continue to provide reliable operational capabilities despite temporary GPS disruptions.

Communications satellites also face substantial hurdles in being targeted effectively. The United States Navy's Naval Telecommunications System (NTS) integrates three primary communication elements: tactical communications within a battle group, long-haul communications between shore-based Naval Communications Stations (NAVCOMSTAs) and deployed units, and strategic communications linking NAVCOMSTAs with National Command Authorities (NCA). The first two elements rely on line-of-sight and extended line-of-sight radio transmissions, covering distances up to 30 kilometers (16 miles) and 500 kilometers (310 miles) respectively, thereby reducing their dependence on satellite infrastructure. Only the strategic communication component is reliant on satellites, meaning that an adversary aiming to disrupt naval operations would need to specifically target these satellites. However, even if such satellites were compromised, battle groups could continue their missions autonomously without direct guidance from the NCA, maintaining operational effectiveness despite the loss of satellite communication.

China's strategic objectives might include attempting to sever communications between deployed units and compelling the NCA to withdraw or stand down a battle group. However, the limitations of ASAT technologies make achieving such objectives exceedingly difficult. The resilience and redundancy built into satellite networks, coupled with the availability of alternative communication and navigation systems, ensure that the impact of ASAT attacks would be limited and temporary. Consequently, while the notion of disabling an adversary's satellite capabilities remains an area of strategic interest, the practical limitations and countermeasures inherent in modern satellite operations significantly constrain the effectiveness of anti-satellite weapons.

About the Author

Thornton D. "TD" Barnes is a distinguished author, entrepreneur, and former military intelligence specialist. Born in Dalhart, Texas and raised on a ranch near Clayton, New Mexico and Dalhart, Texas, he cultivated a passion for exploration. After high school in Oklahoma, Barnes embarked on a ten-year military journey, initially serving in Korea as an intelligence specialist. While in the Army, he also specialized in missile and radar electronics, defending against Soviet threats and later attending the Artillery Officer Candidate School. An injury ended his military career, but Barnes soon transitioned to aerospace endeavors. He worked on significant projects at NASA's High Range in Nevada, including the X-15, the NASA NERVA nuclear rocket project, and atomic bomb testing at the Nevada Test Site. Furthermore, he participated in the CIA's Mach 3 A-12 Project OXCART and stealth projects at Area 51.

Barnes founded and led an oil and gas exploration company outside the aerospace sphere for over 40 years, delving into uranium and gold mining ventures. In retirement, he's dedicated to preserving Area 51's history, serving as president of Roadrunners Internationale and the Nevada Aerospace Hall of Fame Director Emeritus. His contributions have been spotlighted in documentaries on National Geographic, the History Channel, and other major networks. Barnes has authored several books, including "The Secret Genesis of Area 51" and "The CIA Area 51 Chronicles." He currently resides in Henderson, Nevada, continuing to influence aerospace, exploration, and literature, focusing on the formally highly classified of the CIA's era at Area 51.

Bibliography

https://en.wikipedia.org/

Don't miss out!

Visit the website below and you can sign up to receive emails whenever TD Barnes publishes a new book. There's no charge and no obligation.

https://books2read.com/r/B-A-YXRJB-SKTCF

BOOKS 2 READ

Connecting independent readers to independent writers.